Project Management *for* Building Designers *and* Owners

Second Edition

Project Management *for* Building Designers *and* Owners

Second Edition

Howard G. Birnberg

ASSOCIATION FOR PROJECT MANAGERS

CRC Press
Boca Raton   Boston   London   New York   Washington, D.C.

Library of Congress Cataloging-in-Publication Data

Birnberg, Howard G.
    Project management for building designers and owners / by Howard
G. Birnberg. — 2nd ed.
       p. cm.
    Rev. ed. of: Project management for small design firms.
    Includes bibliographical references and index.
    ISBN 0-8493-1265-5
    1. Architectural firms — United States — Management. 2. Engineering
firms — United States — Management. I. Birnberg, Howard G. Project
management for small design firms. II. Title.
NA1996.B5   1998
720'.68 — dc21
                                      98-17157
                                       CIP

No claim to original U.S. Government works
International Standard Book Number 0-8493-1265-5
Library of Congress Card Number 98-17157
Printed in the United States of America 1 2 3 4 5 6 7 8 9 0
Printed on acid-free paper

Foreword

Few activities are as complex as building design and construction. Compounding this complexity is the fragmented nature of the team charged with designing and constructing buildings. In many owner/client organizations, the decision-makers are often unaware of the complexities and costs of the process. The reengineering trend of the early and mid 1990s has greatly reduced the in-house facilities design and construction staff available to plan, coordinate, and manage the owner/client project. Many owner/clients are outsourcing these responsibilities to outside specialists. Some of these specialists are traditional engineering, architectural, and facilities firms. Others are found in newly evolving niches such as program and project management firms.

Traditional design firms are also facing challenges. Specialty firms are newly arrived competitors who have often succeeded in convincing clients of their greater ability to control design and construction costs and to meet construction schedules. Many design firms face other challenges in finding and keeping capable project management staffs, the need for better project scope management, fierce fee and time pressures, and a myriad of other issues.

Outside of the construction industry, the term "project management" often broadly means scheduling. Outside, project management software is scheduling software. Inside, that is only one tiny fragment of the package. However, the entire concept of having project managers is borrowed from outside of our industry, and their fundamental role in managing and controlling the "project" scope, schedule, and budget remains the same both within and without the construction industry.

To achieve their goals, design and owner/client project managers must understand their primary role. They are to serve as the communications conduit in a highly fragmented, specialized, and complex undertaking. Their tools must allow them to communicate quickly and effectively.

The fundamental element of a design firm is its project management system. This system enables firms to complete projects successfully and hence solve the client's problem. The project is the profit center of the organization. The individual who manages the project — the project manager — is in the best position to control the final outcome of a job and can have a great affect on project and firm profitability. However, not all project managers function in the same manner. Many firms have what is known as a weak project management system. Typically, in this type of system the principal or partner is responsible for client contact, while the project manager is responsible for actually producing the work. In effect, the project manager is really a technician.

I have often heard practitioners say that profitability and growth are factors of the ability to "get the work" (success of the marketing effort). I believe that this is only partially true. A much more fundamental factor is the firm's project management system. With a strong project management system, the principals can and should spend more effort "getting the work" *and* managing the firm.

Many firms pride themselves on personal service to the client; however, with a weak or ineffective project management system, the principal often deals with the client and the project manager is in charge of producing the work. No one individual, though, has complete control of the project, hence efficiency and profitability suffer. In essence, when more than one individual is responsible for the project, in reality no one is. As a result, the desired personal service is often a myth because those who are actually responsible for performing the work rarely communicate directly with the client.

Howard Birnberg

The author

Howard G. Birnberg is president of Birnberg & Associates, a Chicago-based firm providing management consulting, association management, and educational and publishing services for the design and construction industry. He is an architect by training, having received his Bachelor's degree in architecture from Ohio State University and his Master's of Business Administration degree from Washington University (St. Louis). Mr. Birnberg is also Executive Director of the Association for Project Managers, an organization of project managers in the construction industry, and is Director of Conferences for the Council on Federal Procurement for Architectural and Engineering Services (COFPAES).

Active in the American Institute of Architects, he served as the general editor for *New Directions in Architectural and Engineering Practice* (McGraw-Hill, 1992) and is the author of *Project Management for Small Design Firms* (McGraw-Hill, 1992).

Acknowledgments

The author would like to thank the following for their assistance and contributions to this book:

Howard Ellegant, AIA, CVS, for Appendix B. Mr. Ellegant is an architect devoted to Value Engineering consulting in the design and construction industry. He may be reached at (847) 491-0115.

Thomas Eyerman, FAIA, for the material on networking in Chapter 11. Mr. Eyerman, former partner in charge of Finance and Administration of Skidmore, Owings & Merrill (1968–1990), is now concentrating upon service to professional firms as an independent advisor and counselor.

Lowell V. Getz, CPA, for the material on contract types in Chapter 14.

Jeffery Lew, for Appendix C. Mr. Lew is on the faculty of Purdue University's Department of Building Construction.

Gene Montgomery, AIA, for the material on project administration in Chapter 14, much of Chapter 17, and all of Chapter 18.

Jeffrey Orlove, AIA, for the section on "Developing a quality assurance program" in Chapter 16. Mr. Orlove is a consultant to design and construction firms. He is the former Executive Vice President and Principal-in-Charge of Management for a major Chicago-based architectural, planning, and interior design firm. He may be reached at (847) 579-1006.

John Schlossman, FAIA, for the material on peer review in Chapter 16. Mr. Schlossman is a retired principal of the Chicago-based architectural firm of Loebl, Schlossman & Hackl. He is a former member of the National AIA Liability Committee.

Contents

Section three. Planning the project

Section four. Managing the project

List of figures and tables

Section one

---

*Organizing for
project management*

Chapter one

The need for project management

Effective project management is important to all design firms. Small firms are especially handicapped by a lack of project control and reporting systems, a lack of staff time solely devoted to managing projects, insufficient principal time to run both the firm and projects, and an inability to market with enough regularity to ensure a steady workload.

In addition, many smaller design practices find it difficult to apply published or seminar material to project management. The situations described in these materials often assume necessary staff and resources to implement prescribed systems and practices. In the typical small firm, principals run projects and the firm. They make all decisions (major and minor), work on the boards, meet with clients, keep project and financial records, negotiate contracts, and are involved in dozens of other tasks on a daily basis. Many of these individuals find it difficult to understand how project management techniques designed for larger firms are relevant to their situation.

There are however, many excellent procedures appropriate for large and small firms that enable firms to escape the treadmill of constant "crisis management". Systems and methods can be implemented that will make design practice more efficient and profitable. For example, firms with shorter duration projects may want to incorporate project management techniques selectively based upon the specific situations presented. It may not, for example, be sensible to establish an elaborate project status reporting system for projects whose duration may be measured in days, not months or even weeks. In this case, the system may not be current enough to allow the principals or project manager to take corrective action in the event of problems. The reporting system should, however, provide enough information to record historical data on project profitability, change orders required, basic contract terms, and similar information.

Table 1.1   Project Management Problems

Problem	Mentions	Percent
High workloads on project managers	14	20.90
Communications problems (internal and external)	10	14.93
Authority needs to equal responsibility	10	14.93
Managing project design and construction budgets	9	13.43
Project manager inexperience and need for training	9	13.43
Finding and training qualified staff	7	10.45
Poor documentation/inadequate reporting systems	6	8.96
Scheduling	6	8.96
Client relations	5	7.46
Motivating and managing the project team	4	5.97
Ineffectiveness of the project management system	4	5.97
Scope changes and management	4	5.97

Project management problems

All firms have their own project management problems. A recent survey by the Association for Project Managers (APM) of 67 design practices and owner/client organizations compiled a list of the most significant project problems affecting them. Dozens of items were listed and some responses were unclear as to the exact nature of the problem. But, there were a number of problems listed repeatedly. Due to multiple answers, the number of mentions shown in Table 1.1 will be greater than the total number of respondents. Percentages shown are percent of total respondents mentioning a particular item.

No design firm or owner organization can ever hope to avoid all of the problems listed in Table 1.1, but a strong project management system can help minimize the impact of many of them.

Design firm life-cycle curve

Few principals objectively evaluate the patterns of growth and stagnation in their firms. Through research data assembled from design firm financial surveys and other sources, a profile of the typical life-cycle of a design practice has been developed (see Figure 1.1). The vertical axis measures common values of growth — typically, fee levels, profits, staff size, etc. Although these are the most common measures, others should not be excluded if they are appropriate for your firm (productivity per staff member, number of offices, etc.).

The horizontal axis measures the number of years of the firm's existence. There are no ideal points or values for each phase, as the

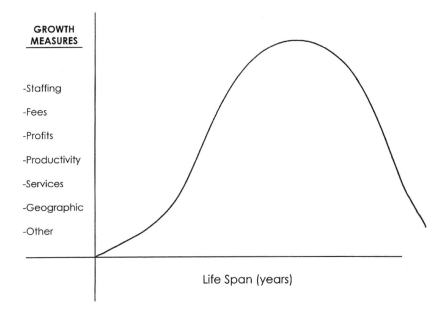

Figure 1.1 Design firm life-cycle curve.

specific levels are only important in that they reflect your ability to adapt as your firm grows, stagnates, or declines. Understanding the curve, the triggers to growth, and the causes of stagnation and decline are the important concepts.

Phase one: organization

Most design firms begin with one or two principals breaking away from an established practice with one or more initial clients. Often, a firm will begin with a 6- to 8-month backlog of work. Unfortunately, as marketing has rarely been a major part of an individual's responsibility at his or her former firm, skills necessary to obtain additional projects are often lacking. Most firms stay small because most principals cannot solve the dilemma of how to do the work and still manage and market. As a result, most principals neglect management and marketing until a crisis occurs, such as running out of work.

A small percentage of the firms grow out of the organizational phase of the life-cycle by early development of fundamental marketing and management systems and procedures. Often, these firms rely on having an interested, effective marketer and a project management system.

Phase two: growth

For a great many firms, rapid growth occurs because of the ability of one or two individuals in the firm to market intuitively. Often, one large project or client will trigger a cycle of rapid expansion. This growth occurs for 4 or 5 years and is followed by a period of temporary or permanent stagnation. Usually, high-growth firms quickly outgrow the systems and processes established during the organizational phase, and attempts to cope with growth often result in patchwork solutions.

For most firms in the growth phase, the following is typical:

- Senior managers' management styles are slow to change even as conditions change.
- Middle managers are often given responsibility without corresponding authority.
- Management and marketing experience is lacking at all levels of the firm's management and is often only obtained after growth has been hindered by poor performance.
- Significant communication problems develop internally and with clients.

To avoid stagnation, systems and management approach must adapt. Senior level managers must devote time and resources to planning, organizing, and marketing.

Phase three: stagnation

The first sign of creeping stagnation is marked by stagnating or declining levels of profit even as the firm increases fees or staff size. Profits decline due to soaring overhead brought on by low productivity and ineffective operations. For some firms, stagnation may initially be caused by the loss of a major client, or by key middle or senior managers leaving the firm to begin their own practices (and repeat many of the same mistakes).

Short-term stagnation can be beneficial as a time of consolidation and reorganization; however, prolonged stagnation (3 to 5 years or more) can be dangerous. Firms deeply into the stagnation phase find themselves losing old clients to other design firms and find old management and marketing systems breaking down or becoming ineffective for a changing marketplace. In some firms, senior principals create a stagnant situation as they grow tired of or disinterested in the practice. Without some change, decline is inevitable. Change can mean selling ownership to young managers, merging with another firm, computerizing effectively, developing aggressive marketing, etc. Prolonged stagnation will be fatal to any business.

Phase four: decline

The decline phase of the life-cycle curve is a direct result of unarrested stagnation brought on by a continued loss of repeat clients and key staff. The decline phase is marked by an extended period of financial losses resulting in severe cash flow difficulties. Bank borrowings often become extensive in an effort to finance continued operations. An aging staff and leadership become increasingly inbred and ineffective.

For many firms, the decline continues as long as a principal remains with the practice. Eventually, a critical point is reached when the firm cannot be saved except by a sell-out or a merger using any remaining assets, contracts, or staff. Often, upon the death, illness, or retirement of the last principal, the firm ceases to exist.

It is impossible to generalize a firm's life-cycle curve from only a few years' experience. There will always be short-term ups and downs due to fluctuations in the economy, sudden loss of a major client, or the illness of a key staff member or manager. However, over a period of years, nearly all firms will experience a pattern similar to the life-cycle curve outlined here. Well-managed firms look for controlled, planned growth; continually conduct long-range planning, marketing planning, and staff and management training; and conduct regular self-analysis sessions.

Chapter two

Project delivery system

Defining the project team

Many organizations and project managers narrowly define a project team. Most consider the project team to include only those immediate firm members working on a particular project. The project team actually should include members of all organizations involved with a project including clients, consultants, suppliers, etc. Each individual team should be represented by a project manager who will act as the permanent representative to the overall project team.

There are as many methods of organizing for project delivery as there are firms. Most well-managed firms are forever tinkering with their systems in an effort to improve service to their clients and their own profitability. Some firms even use different systems within their operation based upon the project size, location, client need, or other factors. However, these systems generally fall into one of three formats.

System types

Pyramid approach

This method revolves around a key individual who makes all major (and often minor) decisions (see Figure 2.1). It is commonly used in smaller firms where the firm's owner has daily involvement in all project and firm management decisions. One or more key technical and administrative support people aid in executing required tasks. Lower level staff members generally perform only assigned tasks. If the key individual is not available, activity often grinds to a halt. In a busy firm, daily crisis management may rule, and important issues, such as short- and long-term marketing, may be neglected.

When a firm has two or more partners, multiple pyramids may exist. Each partner may have his or her own client group and key technical assistant. Formal coordination between partners rarely exists,

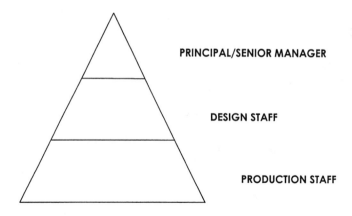

Figure 2.1 Project production system: pyramidal.

and important planning issues are usually neglected. Where one principal is more successful in obtaining work, his or her share of the firm's workload may become overwhelming. This often creates friction and conflict between partners.

Some firms deal with these problems by organizing an informal division of labor whereby assigned project and administrative roles are given to each principal. Unfortunately, most principals still prefer project involvement and continue to neglect their assigned administrative or marketing functions.

Departmental organization

As firms grow, pressure increases to formalize project and non-project management assignments. The division-of-labor concept often becomes the basis for a formal departmental structure. For example, a single-discipline firm often structures itself around marketing, design, production, and field departments. Multi-disciplinary firms may establish each discipline as a separate department (see Figure 2.2).

Single-discipline firms frequently appoint department heads drawn from associate or principal ranks. These individuals normally retain full authority for all project and non-project decision-making within their departments. Responsibility for project management is usually delegated to the next lower staff level, and these individuals are called project managers or project engineers or architects.

The concept behind the departmental approach holds that as one department completes its work on time, within the budget and according to the contractual scope, project authority and responsibility are

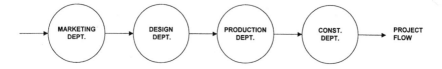

Figure 2.2 Project production system: departmentalization.

passed to the next department head or through the designated manager within the department. In reality, department heads determine project priorities based upon their own workload, level of interest in the project, relationship with the client, and other factors. Often, each department and its head have different methods of project management and capabilities. As a result, the client must adjust his or her operation to three or more different project styles and systems. In addition, if the firm's principals assume departmental responsibilities, firm management and marketing may still be neglected. Under the single-discipline departmental organization, the situation may be little improved over the pyramid approach, as each department may simply become its own pyramid. With no one individual in charge of the entire project, personal service to the client is often a myth, and time schedules, budgets, profits, and the long-term well-being of the firm may all be jeopardized.

In an attempt to deal with the deficiencies of the single-discipline departmental system, some firms have introduced the project manager concept. Unfortunately, this system is not designed or operated properly by many firms. All too often, firms institute a weak project management system where there is a significant imbalance between responsibility and authority. To be successful, levels of responsibility must be generally equal to levels of authority. In many firms, real decision-making authority remains with department heads, while much of the project responsibility is delegated to project managers. As a result, project managers lack the authority, training, or experience to make decisions stick and simply cease being decision-makers. This discredits the project manager/management systems in the eyes of department heads, senior managers, and clients. Where department heads are generally owners and project managers are associates or employees, the situation may become intolerable. In these firms, staff turnover may be high, profitability and productivity low, and client satisfaction questionable.

Multi-discipline firms can use a departmental approach successfully when most projects are within a single discipline. Often, these firms establish a matrix management system within each department. Where studios are used, a similar system may be developed.

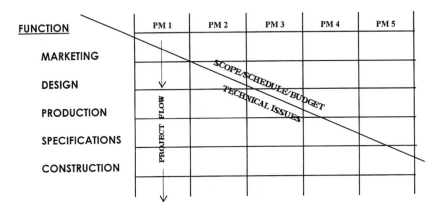

Figure 2.3   Project production system: strong project manager (matrix).

Multi-disciplinary firms with projects involving several disciplines require a project management system to handle work that crosses department lines. Often, a project manager is selected based upon the predominant discipline required. Without an effective project manager, multi-disciplinary firms may find themselves with the same problems as single-discipline, departmentally organized firms.

Matrix management

For many design firms and owner/client organizations, the matrix management (or strong project manager) approach functions best (see Figure 2.3). Under this system the project manager is in full charge of the project from beginning (marketing) to end (continued contact). The project manager has an equal balance of authority and responsibility and is the firm's primary client contact point. His or her major responsibilities include meeting the client's program, schedule, and budget while maintaining profitability for the firm. Department heads or chiefs of the various functional areas (marketing, design, production, etc.) retain responsibility and authority for technical decisions, staff assignments, training in technical areas, implementing quality reviews, etc. Where technical decisions significantly impact the project scope or budget, the project manager (and often the client) must review and approve decisions.

The matrix project management system separates the ownership role from the need to produce a successful and profitable project for both the client and the firm. Unfortunately, at times of conflict, there is a tendency to confuse ownership with a project's real needs.

To be successful, a firm using the matrix management system must be able to communicate to its staff and clients how the system works.

On occasion, confusion can arise on the part of staff members regarding whom to contact with questions, comments, or problems (the project manager or a department/functional head). In addition, matrix management requires a complete, timely, and accurate job-cost reporting system. Other management tools such as a project management manual are also extremely helpful.

Small firms, or large firms with small projects, may establish an abbreviated matrix system. In this case, an individual may wear several "hats" and must be careful to focus on his or her specific responsibilities on a project. Some firms use the full matrix system only on their larger projects and use other approaches on their remaining jobs.

The matrix system requires a strong chief executive to arbitrate disputes between project managers, department/functional heads, and staff members. Although the goal of the system is to avoid bottlenecks and crisis management, disputes will occur that require careful resolution. When functioning properly, matrix systems encourage decision-making at the lowest possible level of the organization, and both responsibility and authority must be delegated to make this possible.

Other concepts

Some organizations use variations of the project delivery systems noted in this chapter. Two systems in particular require some attention.

Account managers

Under this approach, instead of managing projects, a project manager manages a client account. This can be a successful approach, but often leads to imbalances in project manager workload. A particular project manager may become overwhelmed with a very active client, while another may be underutilized by slow client activity or temporary delays in funding, project approval, etc.

Studios

Found in design firms, studios typically are focused on a particular type of work such as health care, industrial/commercial, interior design, etc. Each studio is headed by a senior manager or principal and is run like a semi-autonomous firm. Sharing of staff between studios is not well coordinated and can result in staffing imbalances and higher overhead costs. The benefit of studios is the ability to focus staff and expertise on a specific type of work. Normally a project management system is employed within the studio.

Impact on construction costs

A poorly developed project management system can cost both a design firm and its clients. A design firm's ability to stay within a design fee or to manage design change orders (and often obtain additional fee) is dependent on their project management system. Construction costs can skyrocket when inadequate project management causes poorly prepared construction documents. Contractor bids are either high to cover unknowns or questions on drawings or are "low-balled" to obtain the job (and costs later made up through construction change orders).

Section two

The project manager

Chapter three

Who is a
project manager?

A survey conducted by the Association for Project Managers (APM) shows that design firms and owner/client organizations have vastly different definitions of the role of project managers. Most, however, claim that many or all of their jobs are managed by a project manager (PM); more than 80% indicated that their projects are managed by an individual with the title of PM.

About 58% of the firms surveyed had five project managers or less. The median (midpoint) number of PMs (when including all firms in the survey) was 5.0. Finding and keeping capable project managers remains a vexing problem for many firms. Nearly two thirds (66.1%) of the respondents reported difficulty in hiring and retaining managers. Turnover in the PM ranks is another significant problem, with nearly 48% of firms suffering some loss in the ranks of these key individuals in a 1-year period. Table 3.1 highlights this finding.

In many design firms, principals often assume project management responsibilities. For small practices this is necessary because of the lack of sufficient staff to serve as project managers. In other cases, it is the principal's preference to undertake project management in addition to their other firm management and marketing responsibilities.

Project managers tend to be among the most highly experienced individuals in design firms and owner/client organizations. The mean number of years of professional experience for PMs was 12, and the median was 11.5. The range was from 3 to 26 years. The mean salary for project managers in the APM survey was $53,128, and the median was $50,000.

Table 3.1   Number of PMs Leaving
During the Past Year

Number	Frequency	Percent
0 or blank	34	52.31
1	14	21.54
2	6	9.23
3 or more	11	16.92
Total	65	100.00

Note: Missing = 2.

Characteristics of project managers

The greatest need in many firms is for skilled project managers. Few university curricula provide even rudimentary management training. Necessary project management skills are learned on the job or through seminars or self-initiative. Many designers have little ability or interest in the business side of architecture or engineering. Often, those who show the slightest inclination towards management are pushed to become project managers. In some firms, however, the best technicians are "made" project managers based on their proven ability in a chosen area.

Who should be a project manager? How are they created? What characteristics are important? No one would dispute the fact that the best technicians often make the poorest project managers. The reason for this is obvious. Most individuals with a proven technical ability tend to focus strongly on the aspect of the project of greatest interest to them. With few exceptions, this focus works to the detriment of the broader project needs that a project manager must address. Clearly, it is the best all-round manager that is the ideal project manager. But, what capabilities should this individual possess?

Strong organizational ability

The successful project manager must be able to organize a project and the team and to address the many details that arise. He or she must be strong at organizing manpower schedules and be able to handle more than one major project if necessary.

Generalist

While a manager may have an interest in a particular project area, he or she must be familiar with all aspects of the project. However, project managers do not need to know all of the technical details for themselves. Being an effective delegator is very important.

Insight

To be effective as a project manager, he or she must have a strong ability to examine the broad scope of a client's problem without becoming bogged down in details.

Ability to monitor the project

A project manager must be able to monitor project status and display a willingness to ask for assistance when the situation warrants. Effective communications between members of the project team is vital.

Communicative

Project managers must have good communications skills for both speaking and writing. They must be able to communicate to both individuals and groups as marketers and as managers of the project team. In addition, they must be good listeners.

Presentability

Personal appearance and professionalism must be maintained at a very high level since the project manager is the firm's primary representative to the client. The manager is a public person and must have high regard for the firm's image.

Experience

Successful project managers must have broad experience in a variety of building types. They must have strong skills and experience in project budgeting, negotiating, marketing, and estimating. They should also have their own database of previous project experiences.

Leadership ability

The strong project manager must be a leader, an individual who can direct and motivate their team. He or she should have demonstrated leadership ability prior to becoming a project manager.

Ability to make decisions

A project manager is a decision-maker. The ability to make decisions and to carry them through is vital. In addition, the manager must be able to admit a mistake, and he or she must be able to say no to a client or staff member when necessary. In general, most of the qualities and characteristics of the successful project manager revolve around his or her ability to work well with people, rather than technical skills. Certainly, the PM must have basic technical skills, but overemphasis on these by senior management will not necessarily result in a good project manager. All project managers require regular training to improve skills and to learn new ones. This training must become a regular part of creating project managers.

Communication skills

The key function of a project manager is to communicate. They serve as the primary link between major members of the project team. Each consultant and contractor and the client should be represented by a project manager to communicate their needs, questions, and status to other team members. To accomplish this function, project managers must be skilled communicators. Public speaking skills must be learned and polished through extensive practice or with formal training, such as Toastmasters or a Dale Carnegie program.

Writing skills must be developed to a high degree. Project managers should be sent to university or community college writing courses to learn fundamentals, technical writing, and persuasive writing. Firms should retain, on a full- or part-time basis, an internal staff member to review and criticize written materials constructively and to train all staff in effective written communications.

Chapter four

Project manager responsibilities

One of the most significant changes in design firms and owner/client organizations in recent years is a renewed focus on project management. Gone are the days when any firm could afford poor project management practices. Shorter time frames and tighter budgets are forcing firm managers to make significant changes in their operations. If this is to be the era of the project manager, the following areas will require design firm attention.

Organization

In recent years, many firms have established project management systems. Unfortunately, most of them have a weak version of this system. While project managers may have been appointed, principals and other senior managers are often slow at giving up control over projects. As a result, project managers have titles and responsibilities, but they lack the effective authority to make decisions stick.

For firms to be successful at project management, a true delegation of responsibility and authority must occur. Greater education regarding the role of project management in firms is also required. Principals and senior managers must understand the strong project management system and the role of the individual project manager.

Recruiting

Project management requires a unique set of skills often lacking in technically trained individuals. Many firms lack a clear idea of the role of the project manager. As a result, they are uncertain as to the necessary skills. As was noted in Chapter 3, some firms still select project managers based on technical expertise, not on necessary leadership and management skills.

Many well-managed companies establish career tracks for their employees. This may include a design, technical, or management track. Position descriptions are clearly written, organizational structure is established, and individuals are recruited or promoted to fill a defined need.

Training

Few firms have adequate training programs. This lack of training strongly impacts project managers who require the broadest range of skills of any firm member. A good training program will have several goals:

- Training staff in the firm's methods of operation
- Providing replacements/support for existing staff
- Improving job performance and productivity

For many firms developing a project management system, an individual experienced in project management must be hired to guide the effort. Often, this individual brings an understanding of how the system should function. They may bring forms, checklists, procedures, etc. that can speed up the implementation process. This manager often provides initial training to other project managers. For more on project manager training, see Chapter 6.

Productivity

The years ahead will continue to require enhanced focus on how to improve productivity. CADD and computers can be of great help. Unfortunately, they can significantly change a firm's work methods and can require totally new approaches to producing drawings. Project managers must be aware of these changes or productivity will actually be diminished.

Specialization can affect firm productivity and may require project managers who are experts in a particular type of work. Providing the necessary tools and information to these key individuals will allow them to improve job performance and will enhance productivity.

Responsibilities

There are as many definitions of a project manager's responsibilities as there are firms. In some firms, a project manager is anyone who tells someone else what to do. In one extreme case, a 50-person architectural practice claimed that 20 people were project managers. There is no viable rule of thumb for the number of project managers needed for a certain number of jobs. However, where a full-charge project manager runs the job with the proper tools and staff, a large number of projects can be managed by one individual.

The following is a listing of some of the major responsibilities of PMs. Clearly, this is not a complete list so firms developing a project manager position description should seek additional resources.

Marketing and continued contact

Most experienced clients want to meet and discuss their project with the individual actually responsible for the job. Initial contact and discussions should be held with the marketing staff or principal; however, the project manager should be brought into the process as soon as possible.

The PM should also function as a source of continuing contact with clients on completed projects. This is to ensure the smooth functioning of the facility and to be aware of further services required by the client.

Proposal preparation/fee determination and negotiation

One of the most important responsibilities of project managers is the preparation of a proposed scope of services and the corresponding fee. As the individual responsible for meeting the scope and fee, the PM must buy into them. Scopes and fees imposed on him or her by senior management will not allow for full accountability. With less experienced managers, a thorough review process is essential.

While final contract signing must be done by an officer of the firm, the project manager should lead the fee and scope negotiating process. This provides for full accountability in the event that adjustments in scope or fee are required.

Manpower planning and assembly of the project team

The project manager is closest to the project and the required manpower needs and schedule. He or she must provide regular input to senior management to allow overall firm manpower needs planning.

As the project begins, the PM must suggest names of specific individuals they would like to work on the job. This is based upon their knowledge of the needs of the project and the specific skills of staff members. Obviously, the suggested list must be adjusted based on other projects needs and the availability of various individuals.

Managing the project

The project manager is responsible for supervising all phases of the project, including documenting time charges, meeting budgets (both fee and construction), and handling the many other details of guiding a project through the office. While most project managers should not be making technical or design decisions, in some cases this becomes necessary to ensure compliance with the program.

Quality management

Quality is a shared responsibility; however, the PM must ensure that quality reviews are budgeted for in the project fee and that they take place at the appropriate time. In some firms, where the PMs have the technical competence, they may actually red-line drawings themselves, although this is generally not an appropriate use of their time. It is the responsibility of senior management to develop a quality assurance program. The technical staff and the project managers must ensure that the program is instituted.

Client relations

It is the project manager who can best communicate with the client. There must be a regular process of meetings, telephone calls, and correspondence. The specific and implied needs of the client must be clearly understood by the PM. Managers must control change orders and out-of-scope items by developing a regular process to inform clients of the status of these items.

Project status reporting

A project manager must not only prepare an original project budget, but he or she must also feed the reporting system by providing accurate budgets, contract data, percentages of completion, etc. on a timely basis. Information including revised budgets and maximums on change orders must also be kept current. Project status reports must be closely monitored, and PMs must address problems before they become a crisis.

Billing and collection

Although in most firms invoices are prepared by the accounting office, the project manager should review and approve all invoices prior to issuance to clients. The PM should not be responsible for collecting invoices, but his or her close relationship with the client may expedite collections. For more information on billing and collection, see Chapter 14.

Project managers and marketing

Project managers can and should have a major role in a design firm's marketing effort. Unfortunately, many practices fail to integrate their project managers properly into the marketing team. This failure can seriously affect the marketing success rate and future project profitability. Well-managed organizations train their project managers to have a key function in marketing.

To be effective in the marketing effort, project managers should:

1. *Be assigned to a particular potential project at an early stage.* Once it has been determined that a lead is materializing into a tangible project opportunity, a project manager should be assigned. He or she should participate in subsequent marketing activities and later manage the project if the effort is successful.
2. *Serve as a primary contact person for the future client.* This role is shared with the senior marketer. The project manager must be the client's point of contact on technical and scope issues, while the senior marketer/principal is the liaison on other issues. Clearly, the project manager and marketers must meet regularly and maintain a healthy communication.
3. *Provide technical input to both the marketing staff and the future client.* This role may include assistance in the preparation of a preliminary project program or the suggesting of alternatives that may have lower initial or life-cycle cost, allow for easier expansion, etc. As a result of this effort, firm members may demonstrate their experience, knowledge, and strong concern for the future client's needs and budget.
4. *Participate in the presentation process.* Most experienced clients want to meet the individual who will be responsible for managing and performing the work. The presentation process is an ideal time to involve the project manager with the future client because it allows the project manager to provide specific information as to how the project will be managed. It also helps to establish a good working relationship with the client. The project manager should have an active role in preparing materials and strategies for presentations.
5. *Develop a detailed project scope.* Clearly, the project manager has far more experience than most other marketers in the development of the project scope. With his or her early involvement in

the marketing effort, he or she is very familiar with the future client's stated and implied needs, budget and operational concerns, method of operation, staffing, etc. Project managers are uniquely prepared to outline a proposed scope of services and to evaluate where adjustments in this scope can occur. In addition, it is vital that the individual who will eventually be responsible for delivering a scope of services to the client be involved in the preparation of that scope.

6. *Develop a detailed project budget.* The project manager must, in the development of the project scope, be aware of the costs required to complete the proposed scope of work. He or she must assemble a detailed project budget and outline areas where the budget can be altered by a change in scope or by the negotiations. Without this total understanding of the proposed project budget and scope, the negotiation process will be needlessly complicated. Only when project managers have actually prepared project budgets can they be held accountable for them.

7. *Participate in the negotiation process.* With his or her full understanding of the proposed scope and budget, the project manager is invaluable during final negotiations with the client. Any adjustments that need to be made should be based upon his or her evaluation of the program and upon the ability of the firm to make an adequate fee and profit for its work. No commitments should be made to the client without the project manager's understanding and approval.

8. *Provide input into future manpower and other resource requirements.* A successful marketing effort commits the firm to supplying manpower and other resources to the client's project. The project manager must fully understand these needs and communicate with other managers to ensure their availability. This is an extension of the marketing effort, in that failure to plan for these needs impacts the service provided to the client.

9. *Provide the firm with continued contact with the client.* In all firms, upon completion of a project, personnel and resources are turned to other efforts. As a result, many past clients are inadvertently neglected and little follow-up is conducted on project and building performance. When the client is in need of future professional help, much, if not all, contact with the design firm may have been lost. To prevent this, the project manager must contact the client regularly. The project manager must also be aware of opportunities to suggest to the client changes or improvements in the existing facility (see Figure 20.1).

In addition, project managers should be alert to opportunities to inform clients of the firm's other services. Often, clients have a perception of your firm based upon the services you are currently providing to them. Clients may be unaware of the firm's other capabilities and thus not consider your firm when awarding other projects.

Every staff member has a responsibility to assist in the marketing effort. Project managers, however, are in a unique position to provide job leads, make new contacts, and become involved in community and professional organizations. It is through these efforts that the project manager can also contribute to improving the success of the firm's marketing program.

Survey findings

The effectiveness of project managers can strongly impact project profitability. His or her level of training is an important factor in improving and maintaining effectiveness. Unfortunately, the level of spending by firms on training is inadequate. Significantly, 23 of the 67 (34.3%) firms participating in the APM survey had no documented training expenditures. The mean percentage spent annually on training was 1.49% (median 1.0%).

Table 4.1 (next page) lists the leading responsibilities of project managers as reported by participants in this survey.

Table 4.1 Project Manager Responsibilities

Rank	Responsibility	Number[a]	Percent
1	Consultant meetings/contact	63	94.03
2	Project team meetings	58	86.57
3	Client meetings/contact	55	82.09
4	Change order management	54	80.60
	Project status reports (preparation or review)	54	80.60
6	Personnel planning	52	77.61
7	Scope determination	49	73.13
8	Quality assurance	48	71.64
	Quality control	48	71.64
	Consultant expense review or approval	48	71.64
11	Project budget preparation	47	70.15
	Billing preparation, review, and/or collection	47	70.15
13	Developing a project checklist	45	67.16
	On-site observation	45	67.16
	Construction administration management	45	67.16
	Selection of consultants	45	67.16
17	Project close-out/evaluation	43	64.18
	Proposal preparation	43	64.18
19	Specification preparation/review	42	62.69
	Payment application review and processing	42	62.69
21	Selection of project team members	41	61.19
22	Vendor expense review or approval	40	59.70
23	Submittal review and processing	39	58.21
24	Negotiating designer/owner contracts	36	53.73
	Negotiating designer/consultant contracts	36	53.73
26	Marketing	35	52.24
27	Construction cost estimating	34	50.75
	Construction cost control	34	50.75
29	Partnering process participation	32	47.76
30	Program preparation	29	43.28
31	Project follow-up/post-occupancy evaluation	28	41.79
32	Time-card approval/review	25	37.31
33	Site selection	10	14.93
34	Land acquisition	8	11.94
35	Financing negotiation	6	8.96
36	Other	3	4.48

[a] Total of 67 respondents.

Chapter five

Caring for your project managers

Finding project managers

A continuing source of difficulty for many design firms is finding, recruiting, and keeping capable project managers. As noted in Chapter 3, few engineering or architectural schools teach management skills to any degree. As a result, there is a significant shortage of skilled project managers. Because an increasing number of firms recognizes the value of project management, the competition for available talent is nearing crisis proportions.

In large cities, high job mobility creates the opportunity to recruit project managers from other firms. In many smaller cities, however, the total architectural and engineering community may only number in the hundreds. As a result, experienced managers may be unavailable or cannot be recruited from other local firms or from larger cities. For many firms, there are three basic techniques for obtaining the project management talent required.

Recruit from outside your firm

This method is often the fastest approach to building your management staff. Recruiting from other local firms (particularly in smaller communities) may create animosity on the part of your peers and may also eliminate any hesitancy other firms may have about raiding your own staff. In addition, the local design community may be somewhat inbred, and firms may simply be exchanging each other's weaknesses.

If the local pool of talent is thin, recruiting from other, usually larger cities may be the solution. Unfortunately, attracting staff to smaller communities may not be possible when seeking highly paid, experienced project managers. Offering competitive salaries, fringe benefits, and ownership (or the potential) has been used with varying success.

Train your own project managers

In some communities, the only significant source of project managers may be a firm's own staff. Some firms are reluctant to make a major investment in training their staff for fear of incurring the expense only to lose these people to competing firms after a few years. Clearly, a certain percentage of your staff will leave the firm for various reasons. With sufficient incentive (salary, bonus, ownership, profit sharing, etc.), many capable staff will remain to help the firm prosper. These individuals will have made the cost of training worthwhile.

This training process requires constant budgeting of time and resources for seminars, courses, publications, etc. Some firms recruit prospects directly from colleges and universities to obtain the most capable talent. They then educate these individuals to become project managers compatible with their firm's philosophy.

In large cities, successful firms with experienced teams of project managers also seek younger talent and bring them along as assistant project managers to fill needed slots. In many locations, it is not unusual to find a large percentage of design professionals who have worked for one or two local firms at one time. Many of these firms are noted for their training programs.

Recruit an experienced project manager as the mainstay of your staff

For many firms not experienced with effective project management, it is often wise to recruit one knowledgeable manager as the center of your system. This individual should help establish the project management program, recruit and train younger staff, and serve as a technical and managerial resource. In many firms it is not necessary to recruit an experienced manager, as a principal may wish to begin an intensive self-education program to acquire the necessary skills.

Keeping your managers

Finding and training your project managers is only the first step. Keeping your hard-won managers is just as important. It is the responsibility of senior management to provide for the psychological and financial well-being of these individuals. The obvious incentives of competitive salaries, bonuses, profit sharing, and a fringe-benefit package are most important. As noted in earlier chapters, project managers must have authority that matches their level of responsibility in the firm. Second-guessing and countermanding their decisions will quickly destroy your project management system, and many of your managers may become interested in opportunities with other firms.

How many project managers do you need?

A frequently asked question is how many project managers does a firm need? Unfortunately, there is no fixed answer for a number or reasons:

1. *The experience level of your project managers is an important factor.* Generally, more experienced PMs should be able to handle more projects. They should also be capable of managing projects of greater complexity than managers with less experience. In many cases, firms employing less experienced managers may require a greater number of individuals to handle the workload.
2. *The experience level of your technical staff is a factor.* Staff experienced in a particular project type can make the PM's job a great deal easier. Their familiarity with the workings of the firm's project management system is also important. Every employee knowing how to best perform his or her job can improve efficiency and communications. This allows project managers to focus on important activities.
3. *The quality of your support and management information systems will impact the ability of project managers to perform.* Effective project management requires a wide array of tools and systems. When managers lack all or part of these systems, more time is required to complete their project responsibilities. This may mean the firm will need a greater number of PMs.
4. *The complexity of the specific project a manager handles will affect his or her ability to manage additional projects.* Managing complicated medical or manufacturing facilities will require greater effort and attention than basic speculative office buildings or retail spaces. Increased time demands may be placed on a project manager to administer a complex project.
5. *The geographic location of projects will impact the manager's time.* Projects located in distant or remote locations will involve more travel time for client meetings, job site visits, and general coordination.
6. *Experienced and sophisticated clients can ease the burden on your project managers.* The smooth functioning relationship between the project

manager and an experienced client can allow a PM to handle additional projects. Occasionally, the sophisticated client can actually require more of a project manager's time by demanding additional services or attention. However, this may be preferred to the "hand-holding " required with an inexperienced client.

7. *A major factor determining the number of required project managers is the general staff experience level with the current projects of the firm.* When a one-of-a-kind, rarely handled project type is encountered, the learning curve is higher for all. Increased levels of research will be required as will increased levels of client interaction. This will usually necessitate greater time commitments for the PM.

8. *The current number of projects under contract is an important factor.* Most project managers can handle several projects at one time. An increased number of jobs in the office either will result in a higher work load per PM or will require the hiring or training of additional project managers.

9. *The mix of projects normally handled by the firm can affect the number of project managers.* For example, an office with a large number of small projects will likely need more PMs than one with a few large jobs. Because most projects require at least a minimum level of service, more jobs mean a higher work load.

10. *The timing of your projects can put extraordinary demands on project managers.* Most firms strive to have PMs handling projects in differing phases to avoid "crunches". Typically, one project may be in a start-up phase, another heavily into design or construction documents. A third may be in construction administration. Unfortunately, delays and changing programs may alter this desired scheme. At those times, the pressure on project managers to perform can be extreme.

With no easy way to determine the required number of project managers, how do firms cope? Well-managed design practices are always training younger staff in the principles and applications of project management. Junior staff can assist experienced PMs to learn project management. After a little seasoning, they can be given the opportunity to manage their own smaller, less complicated projects. In this way, a cadre of people is always available to meet the needs of an ever-changing workload. For more information, see the cross-training discussion in Chapter 6.

In small firms with limited staff, all employees should be given basic project management training. It is vital that the required tools and systems be in place to assist in managing projects.

Rewarding project managers

Senior managers in many firms find it difficult to determine appropriate rewards for project managers. As key employees of the firm, project managers deserve a high level of base compensation. In most parts of the country, project manager salaries begin at no less than the mid-30s and can range upwards of $80,000 in major cities and in large firms.

Beyond base salary, what are the best methods to reward project managers? Smart firm managers generally try to assign their best project managers to the most difficult projects. This may be the most technically complicated job, one that has a tight time schedule or a very restrictive fee budget. Basing a project manager's rewards upon the project's eventual profit level will be extremely unfair to the good performer assigned to a tough job.

Some firms have attempted to base both the project manager's reward and that of the entire project team on the final profit level of the job. During the fee budgeting process, senior management and the project manager set a target profit level for the project. It is agreed that any profits beyond that level will be divided among the members of the project team, with the manager receiving as much as half.

Short-term gains may be achieved by profit sharing at the expense of other projects and the entire firm. Obviously, if I stand to gain financially from a project making a substantial profit, I will put forth a significant effort. I will also choose carefully where to focus my efforts. A project with a lower profit potential due to size, complexity, or other factors will likely receive less of my attention.

Some try to make this work by providing a "down-side" risk to project managers. They are penalized for projects that lose money and this loss is deducted from the profitable projects. However, if losses exceed profits, the system nets out to zero and no further penalty is incurred. Few project managers would bet their salary against losses.

Project managers typically do not have control over resources, such as the assignment of team members, schedules, and, often, the selection of consultants. It is these resources that can make the difference between a profit and a loss. Sharing project profit with project managers and team members can encourage short cuts, poor client service, and destructive internal competition.

Positive rewards

Firms will want to provide an atmosphere conducive to positive self-esteem. Project managers must not only have a high level of responsibility, but must also have a corresponding level of authority. Senior management must trust the judgment of project managers and publicly back their decisions, even if private follow-up is required.

Project managers must also have significant input into decisions for which they will later be held accountable. For example, a project fee budget should be prepared by the project manager and reviewed by senior management, not the reverse. This process rewards and encourages the project manager by providing a positive environment of trust and confidence.

A spot bonus program is an ideal way to reward good performers and should be used by all firms. While the amount of money is not significant, the symbolism is very important. Salary increases and

---

Project manager manuals

Development

Among the most important training and management tools for project managers is the project management manual. Unfortunately, very few firms prepare this valuable document. Its general purpose is to document most of the responsibilities and tools required to perform the project management role in a firm.

The manual should be in a loose-leaf format to facilitate its use and to allow for easy revisions. Each section should provide for a "sunset" date, at which time it is to be revised. The manual preparation should be managed by a senior executive, while responsibility for research and writing is assigned to several project managers. The sections must be assigned like a project, with specified due dates and content. A professional editor/writer should prepare the final draft.

Typically, the manual is used as a reference document by experienced project managers, as a training tool for new or less experienced managers, and as an orientation tool for new staff members at the middle and senior levels of the firm. Often, firms provide copies of all or part of their manual to their clients in an effort to inform them of the firm's procedures and methods of operation. This can clarify critical issues in the early stage of a project.

Content

In general, a project management manual should include the following:

1. *The firm's project management philosophy and approach.* This is a brief section on the firm's concept of project management and organization.
2. *The role of the project manager in marketing.* This must include an outline of the firm's marketing plan and structure and a clear explanation of the expected interface of the project manager and the marketing.

promotions are obvious rewards for good long-term performance, but the use of increased fringe benefits can also be a valuable reward. This might include providing a company car or other similar reward. Firms should also have profit sharing or similar programs in place to reward all staff members for superior year-long performance.

Salaries

The salaries of project managers vary greatly depending on the office location, responsibilities, experience, and many other factors. In many firms, project managers are the second highest paid individuals after

---

3. *Project front-end planning and contractual activities.* This section clarifies the project manager's role in preparing the project program and budget and outlines responsibilities and work assignments. In addition, the role of the manager in the writing and negotiating of contracts must be reviewed, and copies of all required forms should be included for illustration purposes.

4. *Management of the project team.* Responsibilities for meetings, communications, personnel issues, and performance appraisals are outlined in detail. In addition, consultant and client-relation activities (including the handling of out-of-scope items) are discussed, and copies of all current forms are included. Detailed project checklists and position descriptions for most project roles are also essential.

5. *Quality management.* Although project managers generally do not conduct quality reviews themselves, they are responsible for ensuring that a time and budget allocation exists for this activity. They must also require that quality assurance reviews are conducted as scheduled.

6. *Control of budgets and schedules.* Requirements for review of budgets and project status reports must be outlined, and standard forms should be included. Standard operating procedures for the administration of these items (particularly change orders) are outlined here. A procedure for corrective actions in the event of problems should be included.

7. *Project close-out and follow-up activities.* Checklists and procedures for completing a project are reviewed in this section, which also includes samples of all required forms. Instructions for project follow-up and continued contact must be outlined.

Although enhancements to the manual can be included, these sections provide the basic information required. The key to a successful project management manual lies in its completeness and ease of use. Any system or manual that becomes too complex or cumbersome will not be used.

Table 5.1   Average Salary of Non-Principal
Project Managers

Range (×1000)	Frequency	Percent
$30.0 – $39.9	3	5.66
$40.0 – $49.9	18	33.96
$50.0 – $59.9	18	33.96
$60.0 – $69.9	10	18.87
$70.0 and higher	4	7.55
Total	53	100.00

Note: Missing = 4.

the principals. Table 5.1 shows the average salaries of non-principal project managers as determined by a survey conducted by the Association for Project Managers (APM). The mean salary for non-principal project managers was $53,128, and the median was $50,000. The range was from $32,000 to $110,000.

Chapter six

Training project managers

Few members of your staff are of greater importance to your success than the project managers. Their pivotal role among clients, contractors, and staff requires them to possess a unique set of skills. Unfortunately, many project managers are forced to learn on the job. The benefits to them and the firm of a formal training program are great. What should a project manager training program include?

Project manager training programs

There are three broad areas that should be covered: communication skills, interpersonal skills, and technical management skills (see Figure 6.1 for a suggested list of training topics for project managers).

Communication skills

Project managers need to possess a wide range of communication skills; their importance to the marketing effort is well documented. Well-managed firms seek to involve these individuals at a very early stage when contacting a potential client. As a result, their experience and skill at marketing and selling are essential.

Other communications skills are also vital to project managers — in particular, negotiating, effective writing, and public speaking. A project manager's involvement in negotiating contracts and with other members of the project team such as consultants makes this an obvious area of focus. There are a number of commercially available negotiating courses.

Project management concepts/systems
Quality assurance/Total Quality Management (TQM)
Computers (hardware and software)
- Scheduling
- Estimating
- Budgeting
- Database systems
- Project status reports
- Presentation software
- Proprietary systems
- CADD
- Internet/research

Contracts/risk management
Interpersonal/communication skills
- Writing
- Public speaking/presentations
- Delegation/motivation
- People skills/managing people
- Negotiation
- Working with others

Construction inspection
Time management
Financial management
Project budgeting
Scope management

Figure 6.1 Suggested project manager training topics.

Many design and construction staff are poor writers. Much of their writing suffers from wordiness, improper punctuation, capitalization, run-on sentences, and a long list of other grammatical faults. This inability to write effectively and properly reflects poorly on your firm. Many local community colleges and universities (such as the University of Wisconsin Engineering Professional Development) periodically offer effective writing courses.

Public-speaking opportunities for project managers include community groups, social organizations, client meetings, project team meetings, and marketing situations. Public speaking is high on most individuals' lists of major fears. This fear is often only overcome through practice. While some project managers practice presentations before project teams or family members, others seek more directed and instructive environments offered by groups such as Toastmasters, which has local affiliates in nearly every major U.S. city.

Interpersonal skills

Project managers are people managers. They must know how to direct, motivate, and manage their project team, contractors, clients, suppliers, and many other individuals with whom they interact. For some, this is a natural ability; for others, it requires extensive training in human psychology. Numerous sources exist that can help project managers improve their ability to work with people.

A difficult skill to teach is leadership. Some individuals exhibit natural leadership skills, but those who do not can learn techniques to improve their leadership ability.

Perhaps the most difficult skill to learn is that of delegation. Many design professionals tend to be poor delegators, are ego driven, and often lack trust in their subordinates' skills. As a result, these individuals feel the need to be involved in all aspects of the project at all times. Not only does this overburden them, but it also hinders the performance of project team members. Learning how to delegate is a painstaking process that must be reinforced by the example of top management and the provision of tools and systems that permit adequate supervision of subordinates.

Technical management skills

To be effective, a project manager must have a complete understanding of technical management skills. This covers a broad range of project activities. Project budgeting, scope determination, manpower planning, and quality assurance reviews are only a few of these tasks.

There are a number of outside sources for assistance in developing or enhancing these capabilities — in particular, the annual week-long course (held in early January) offered by the University of Wisconsin-Engineering Professional Development and programs offered by a number of local affiliates of the national professional societies. In addition, several excellent texts are available on these topics (see the Bibliography for more information).

Developing your training program

Most firm managers are oblivious to the how's and why's of an ongoing training program. This continuing education of staff and management requires a commitment, a plan, and a budget. Unfortunately, most firms leave this process up to each individual, clearly subjecting the firm's future to chance.

Farsighted firm managers offer opportunities for staff and management to learn or improve their skills. Methods vary, ranging from in-house seminars to paid tuition at local colleges. Training not only improves skills, but also serves as a morale booster and a fringe benefit while protecting the firm's future. Employee turnover is often the rationalization for not providing formal training. "Why train someone else's staff at our expense?" is often the philosophy.

A recent study found that 75% of high-technology companies experienced a significant increase in employee productivity after developing and funding comprehensive formal training programs.

Experience indicates similar results for architects and engineers. Clearly the cost of training is more than matched by productivity increases and is favorable towards training. In general, training goals fall into three categories:

1. Teaching employees/managers how to perform a new or unfamiliar task within their current job
2. Helping employees/managers improve their performance on their present job
3. Preparing employees/managers to handle new jobs

Staff/management training

There are two major areas of training for an organization: staff and managerial. Staff training usually consists of enhancing specific skills such as drafting, CADD, product/service knowledge, etc. Training in these areas is usually very direct, observable, and objective and can be broken down into a number of discrete parts or elements.

Managerial training, however, usually focuses on communication skills, supervisory skills, human relations, etc. Although these types of skills are more subjective and harder to quantify and measure, they may have a greater impact on your organization. A firm may have the finest, most talented engineers or architects in the country, but without proper supervision, direction, and motivation, this talent may be unproductive.

Determining if training is needed

Often training is conducted for fairly limited reasons. These include teaching new skills to recent or current employees, re-training employees in skills they may have lost or not used in many years, or keeping employees abreast of changes in technology and design, etc.

Your first step in the training process is to determine the need for training. A thorough needs analysis should be performed on the organization according to employee and position. In assessing the organization's needs, it is necessary to look at the firm as a whole. What are its strengths and weaknesses, and how does it compare to its competition?

In determining the firm's overall needs, it is necessary to look at both short- and long-term needs. For example, if a senior partner who has handled most of the firm's marketing will be retiring in 2 years, now is the time to start training someone to take his place. It is important at this level of assessment to consider the short- and long-term goals of the firm. These will have an impact on what the training needs are or will be.

Another important aspect of this organizational assessment is the "climate" of the firm. The firm's attitude and motivational level will have a great impact on the success of any training programs that are instituted. One method of assessing the general training needs and attitude of staff is to interview and have questionnaires completed by senior management, project managers, etc.

The second level of assessment, along position lines, will help determine more specifically what and where training is needed. A thorough analysis must be done on each position in the firm to determine not only the duties and responsibilities of the position, but also the skills necessary for a person to do the job successfully.

This type of assessment involves a formal, systematic study of a position that covers a number of items. This includes what persons in each position do in relation to information or other people; the procedures and techniques they use; the equipment, tools, machinery, etc. they need; the products or services that result from their effort; and the skills, traits, and attributes required of the person in the position.

Finally, an analysis should be done on each employee and senior manager to determine the skills each person has or lacks. This will help determine the training they may need to better perform their job, what position they can move into next, and what job they could grow into in the future. Assessment of employee skills may involve reviewing performance evaluations, reviewing their work, completing questionnaires, and conducting skills or ability tests.

When looked at as a whole, the identification of training needs is an involved and complex procedure. It will, however, allow a firm to assess its strengths and weaknesses, focus attention where needed, and grow in the direction desired.

Few firms have a well-established training program. Most simply take advantage of isolated seminars, and often only senior management attends these programs. Training is often considered the responsibility of the individual, who is expected to plan, schedule, and finance his own program. As a result, most firms are not adequately prepared to respond to the need for new services or to meet changing market conditions.

A staff training program requires a long-term commitment and a recognition that the payback may not be immediate. Regular training will result in a more productive and profitable firm. How should a training program be developed?

1. *Commit to a continuing program.* A program that is conducted on an irregular basis will never achieve its goals.
2. *Establish an educational planning group/staff development task force composed of three or four individuals representing all staff levels and chaired by a principal.* This group should be charged with developing and managing the training program, researching training options and techniques, and preparing specific programs. They should meet regularly (at least once a month) and should operate on a priority basis. The Staff Development Officer (see sidebar) should be a member of the task force.
3. *Develop a training plan and schedule.* This should include choosing various types of training, establishing training priorities and goals, and outlining who is eligible for each program. In addition, a schedule should be established to guide the training process.
4. *Establish a training budget as part of the annual budgeting process conducted by the firm.* These are funds that should be spent and

---

Staff development officer

Some organizations hire a staff development officer to manage the personnel operations of the firm and to administer the training/education program. This individual will

- Assure compliance with all legal issues regarding personnel, including state licensing requirements for continuing education
- Assist the chief operating officer in the development of staff benefit/reward programs
- Maintain personnel records
- Assist in the hiring/departing (voluntary/involuntary) of staff
- Administer records related to staff training/education (including CEUs, state licensing, etc.)
- Research training/education needs and opportunities
- Contract for training/education
- Prepare a training/education opportunities bulletin to be distributed to all staff on a regular basis

not viewed as an area to cut if the firm experiences temporary declines in workload.

5. *Inform your staff of the various training options available and what items the firm will pay for.*

6. *Require those attending outside educational programs to disseminate their information to other staff members.* This could be done at lunch meetings where short presentations are made or in a summary report on the program.

7. *Vary the types of training programs used.* Many options are available, including:

   - In-house lectures and seminars — these programs may last from 1 hour to 1 day and may be conducted by outside management consultants and specialty consultants, building product manufacturers, college professors, or experienced and knowledgeable staff members.
   - College courses — tuition may be paid in part or in full for certain staff members to expand present capabilities or develop new ones. Correspondence courses should also be considered.
   - Outside seminars — the National Society of Professional Engineers (NSPE), American Consulting Engineers Council (ACEC), and many other organizations sponsor numerous

---

   - Serve as a clearinghouse for all solicited and unsolicited material on training and educational opportunities
   - Help each staff member to determine their individual training/educational needs
   - Assist in the development of an individualized training program for each staff member
   - Serve as a member of the staff development task force
   - Develop an application for training form and administer the review process
   - Prepare an annual staff development budget and manage expenditures
   - Develop and maintain a library of personnel, training, and educational resource materials
   - Administer the tuition reimbursement program
   - Help to determine incentives for each staff member to pursue training and education
   - Assist in the development of career paths for staff
   - Help in the creation and implementation of a mentoring program

part-day, full-day, or multi-day courses and seminars in major cities. Many universities (for example, the University of Wisconsin, Penn State, and Harvard University) regularly offer short seminar courses. Although the cost of attending many of these programs is high, they give staff and principals the opportunity to exchange information and ideas with individuals from firms throughout the country.

- Professional conference/conventions — these may include conventions and conferences organized by professional design groups (AIA, ACEC, NSPE, etc.), suppliers, product manufacturers, client groups (American Hospital Association, etc.), and others.
- Audio and video tapes — although many of these tapes are too short, poorly produced, and expensive, some may have lasting value as reference and refresher tools.
- Resource materials — an important part of a good training program is a library of reference books, magazines, etc. This material must be organized into a usable collection that is regularly maintained and updated.

8. *Performance review of your training program.* At least once a year, the entire program should be reviewed for its effectiveness, cost, and impact on morale and productivity. The budget must be evaluated for its short- and long-term cost effectiveness.

What makes for effective learning?

Probably the most critical factor determining the success of training is the motivation and attitude of the people being trained. The trainees should want to be trained and should believe that the training will have positive results. A number of steps can be taken to help foster these feelings, before and during the training process, the most important being that the goals and desired outcomes are clearly conveyed to the trainees.

During training sessions, there should be rewards for learning the material. Reinforcement must be provided for making use of what was learned and for proper learning or training behavior. Trainees will learn and remember material that they consider to be meaningful and important.

Every training program should begin with an overview of the material to be covered and an explanation of how it relates back to job problems or performance. The material should be broken into logical pieces, and these should be put into a rational, progressive sequence. The terms used and the technology discussed should be familiar to the trainees. New terms and technology should be presented in a manner that the trainees can relate to. Visual aids should be used whenever possible.

Cross-training

Design and construction organizations must provide cross-training for all staff members. There are three primary benefits to cross-training:

1. Staff develops the necessary skills to be used in the event of turnover, vacations, illness, etc. It is vital to maintain continuity of service to clients. Cross-trained staff can more easily maintain your high service standards.
2. Expanding workloads will require additional trained staff and project managers to service new and existing clients.
3. Cross-training allows everyone to perform their jobs better. Understanding why information must be in a particular form, how work is to be completed, team members' information requirements, etc. helps to improve everyone's performance.

Training practice and methods

There are a number of different methods by which training materials can be presented. The method of presentation will impact the effectiveness of the program. No single method can be used for all types of material. A training program should be designed for maximum efficiency within the constraints of time, cost, location, equipment availability, etc.

Lecture method

Probably the most familiar and widely used instructional method is the lecture. It is usually done live, but it may be presented on video or audio tape. From a training/learning standpoint, the lecture is one of the weakest methods. It usually involves no interaction, practice, study, or testing of the material presented.

Classroom training

Traditional classroom training is essentially a series of lectures. Classroom training allows for modifications and enhancements to the lecture method by providing workbooks, small group discussions and practice, multiple sessions with homework, regular testing, etc. This approach is certainly more effective than a simple lecture and is appropriate for more complicated training. In a firm, it could be adopted for almost any type of staff training. This could include specific technical skills, such as drafting, drawing, CADD, etc. This is especially true when enhanced with a "lab" set-up, using CADD terminals or drafting for hands-on practice.

Programmed learning

Probably the best method, at least for skills training, is programmed learning. This method breaks the training program down into many smaller parts that are put into a logical sequence. At the end of each section or module, the participants are tested and given immediate feedback as to their understanding of the material. Training courses of this type are much more difficult and costly to develop and usually involve programmed texts or workbooks. There are many advantages to this type of training approach. It is designed to be individually paced, with each participant moving at his or her own speed. Frequent testing can determine if the material is being learned.

Group discussion

This method is very familiar to most professionals and can be used as a separate approach in and of itself or in conjunction with lectures. In this method, small groups discuss issues or problems, work out new ideas, solutions, proposals, etc. It is most effective for teaching problem-solving and decision-making skills, presenting complicated or difficult material, or changing opinions and attitudes. As such, it is probably most useful for management, rather than staff-level training. It is particularly useful in human relations, communications, and supervisor training programs. Depending on the nature of the material covered, it may include intense confrontation and discussion or argument, role playing, case studies, management "games", simulation exercises, etc.

This material is only an overview of training program approaches. All of these programs can be used either in-house or at outside locations but are typically conducted "off the job" or "off-site". These approaches take the trainees away from their regular jobs. In many cases, programs away from the work site will result in a more productive learning atmosphere.

On-the-job training

Another type of training method is on-the-job training (OJT). This is often used to train for specific skills. Trainees learn while they are actually on the job and are being productive. OJT is usually combined with classroom training or other off-the-job approaches, as well. Internships are an example of this approach. In a design firm, the OJT approach may be used for drafting, CADD, design work, etc. Firms often combine the OJT method with other approaches for better results.

Mentoring programs

Mentoring programs match an experienced individual (the mentor) with a less experienced individual (the novice). The mentor is to guide in the development of the novice's skills and career path. Mentoring programs fall into two categories: formal and informal. Under a formal program, a structure is established to guide both parties. Career paths should be in place. Regular meetings are held, tasks are assigned, goals are developed, evaluations are held, and a wide variety of similar tasks occur. Under an informal program, the parties simply develop a working relationship where the mentor takes an interest in the development of the novice.

Program evaluation

The purpose of a training program is to increase employee and organizational performance and productivity. The evaluation of any training should therefore focus on measuring these factors.

Managing the training program

A decision must be made whether to develop and provide training in-house or purchase programs from outside vendors. There are advantages and disadvantages to each approach. A primary consideration in deciding to start in-house is whether or not you have the expertise and the facilities to do so.

Another major consideration is your budget. There is a cost to providing training. As with any other project, costs must be determined and a budget established. The overall budget must cover a number of items. These include training materials and supplies, facility use/rental, instructors' salaries, price/cost per trainee, and the loss of productivity while the trainees are off the job. As a rule, a design firm should spend at least 5% of its annual total revenues on training.

Consideration must be given to a program's timing, as well to determining which month, week, or days of the week are most convenient. The location and facility must be chosen. Participants, supervisors, and managers must be notified of all details.

Some of the above steps can be eliminated if a decision is made to go with an outside training provider. However, careful effort should be put into evaluating and choosing an outside provider.

Sources/providers

There are literally thousands of outside providers of training programs. Most, if not all, professional organizations (such as the AIA) either conduct or sponsor professional training. The American Society for Training and Development (ASTD) is an organization that would be a valuable contact. The same is true for the American Society for Personnel Administration (ASPA; see the Bibliography for more information on these organizations). Other providers of training services are management consulting firms, industrial psychologists, colleges, universities, and other schools. Numerous books have also been published on the subject of training. Check with bookstores or your local and university libraries for titles on this subject.

Training of young staff should begin immediately upon their graduation. The AIA has developed an excellent method called the Intern Development Program (IDP). This program provides a structured framework that exposes young, unlicensed architects to the specific areas of practice required to pass state licensing exams and eventually contribute to their employers' practices. The IDP also provides for special advisors and offers a series of study guides covering all areas of architectural practice.

Section three

Planning the project

chapter seven

Profit plan

In a national survey of design firms conducted by Birnberg & Associates, one question addressed the issue of the preparation of annual profit plans. Out of 152 firms responding, 108 (71%) prepared such plans. Many of these firms were large, successful companies. Unfortunately, many smaller firms fail to follow suit.

The definition of a profit plan is a "management tool for formalizing the firm's financial objectives." The benefits of preparing a plan are many, including:

1. Establishing yearly goals
2. Providing intermediate targets throughout the year

The remainder of this chapter will discuss how to prepare a profit plan and apply it.

Figure 7.1 provides a blank form which you may copy for your immediate use. Figures 7.2 through 7.5 isolate portions of the plan for discussion. Note that consultant and non-consultant reimbursables and the markup on reimbursables are not considered on the profit plan. The profit plan is prepared in the same manner whether you manage your firm on a cash or accrual basis (not your tax basis).

Labor

Figure 7.2 isolates the portion of the profit plan concerned with labor allocation. Labor for the firm has two components: project chargeable (direct expense) and non-project chargeable (overhead). Each principal's time is analyzed based on historical records and future projections of workload for the breakdown of project chargeable vs. non-project chargeable. An overall average for all principals is calculated and entered as in the example (60% project assigned and 40% unassigned). The importance of complete and accurate time sheets is readily seen.

	Total ($)	Project expenses ($)	Overhead ($)	Profit ($)
Principal's draw	____			
____% project assigned		____		
____% unassigned			____	
Technical salaries	____			
____% project assigned		____		
____% unassigned			____	
Administrative salaries	____			
____% project assigned		____		
____% unassigned			____	
Total labor	____	____	____	
Non-reimbursables	____	____		
Overhead	____		____	
Net profit (before tax)	____			____
Net fees	____	____	____	____

Income statement		Target	
		Percent	Time-card ratio
Net fees	$____	____	____
Labor	($____)	____	____
Non-reimbursables	($____)	____	____
Gross profit	$____	____	____
Overhead	($____)	____	____
Net profit (before tax)	$____	____	____

Figure 7.1 Profit plan. (Adapted from Bevis, Douglas A., *Profit: Planning For It, Making It, and Keeping It,* Naramore, Bain, Brady & Johanson, Seattle, WA, 1976.)

A total figure for principal's draw is prepared (exclusive of bonuses, profit sharing, etc.). As shown in Figure 7.2, this amount is $150,000 for the year. Hence:

$150,000 × 60% = $90,000 direct expense (project assigned)

$150,000 × 40% = $60,000 overhead (unassigned)

	Total ($)	Project expenses ($)	Overhead ($)
Principal's draw	150,000		
60% project assigned		90,000	
40% unassigned			60,000
Technical salaries	350,000		
83% project assigned		290,000	
17% unassigned			60,000
Administrative salaries	100,000		
20% project assigned		20,000	
80%unassigned			80,000
Total labor	600,000	400,000	200,000

Note: All figures are for illustration purposes only and should not be used as targets for your firm.

Figure 7.2 Sample profit plan (labor only).

This same process is repeated for the technical staff and for administrative staff. The 20% project assigned for the administrative staff often comes from typing of specifications, reports, etc. by secretaries. Each column is totaled to achieve the total labor line. The following are required to complete this section:

1. Total salaries and a breakdown by principals (or partners), technical staff, and administrative staff projected for the coming year
2. Estimated chargeable rates for each principal and staff member projected for the coming year

Non-labor costs

Figure 7.3 adds all other costs (other than labor) to the plan. Non-reimbursable direct expenses are project chargeable expenses that come out of your fee. This would include items such as printing, entertainment, travel, some consultants, and other similar expenses. As these expenses are all project related, they are listed under direct expenses. In general, you should seek to minimize these expenses.

Overhead totals are established by performing a detailed analysis of projected expenses (including all fringe benefits) for the coming year.

	Total ($)	Project expenses ($)	Overhead ($)	Profit ($)
Principal's draw	150,000			
60% project assigned		90,000		
40% unassigned			60,000	
Technical salaries	350,000			
83% project assigned		290,000		
17% unassigned			60,000	
Administrative salaries	100,000			
20% project assigned		20,000		
80% unassigned			80,000	
Total labor	600,000	400,000	200,000	
Non-reimbursables	10,000	10,000		
Overhead	400,000		400,000	
Net profit (before tax)	390,000			390,000
Net fees (revenues)	1,400,000	410,000	600,000	390,000

Figure 7.3 Sample profit plan and non-labor costs.

In Figure 7.3, it is estimated that overhead for the coming year will be $400,000, which is listed in the Overhead column.

Net profit (before tax) is determined by the establishment of attainable profit goals by the firm's principals. For partnerships, this figure would be the total of the partners' desired shares in addition to salary draw and any planned distribution to the staff. A reasonable profit target is 17% to 20% of total revenues (without reimbursables) or 25% of net revenues. The example shown in Figure 7.3 indicates a target net profit of 27.9%:

$$\$390,000/\$1,400,000 = 27.9\%$$

Based upon the profit plan, the firm has targeted $1.4 million in net revenues for the year. Net revenues are those earned based upon your efforts only and do not include any pass-through items, such as general reimbursables and consultant reimbursables. In the event that the economy or other factors will not permit this level of revenues, a rebudgeting should be performed. Remember, this is a worksheet and should be recalculated based upon new information and/or conditions.

Ratios/multipliers

Net revenues (total or gross revenue less non-reimbursables and consultants) are the basis for calculation at 100%, and each line item is
converted to a percentage of revenues:

$$\$400,000 \text{ (direct labor)}/\$1,400,000 \text{ (net revenues)} = 28.6\%$$

Each item is calculated in the same manner.

A time-card ratio or multiplier is calculated by using direct labor
(i.e., total raw labor without fringes) as a base of 1, as shown. All other
items are expressed as a factor of direct labor — for example:

$$\text{net revenues}/\text{direct labor} = \$1,400,000/\$400,000 = 3.5$$

$$\text{total overhead}/\text{direct labor} = \$600,000/\$400,000 = 1.5$$

$$\text{non-reimbursables}/\text{direct labor} = \$10,000/\$400,000 = .025$$

$$\text{net profit}/\text{direct labor} = \$390,000/\$400,000 = 0.975$$

If a firm's 3.5 multiplier (based on direct raw labor without fringes) is
too high for the local market conditions, then overhead or other factors
can be adjusted to reduce the multiplier. (Once the multiplier is changed,
recalculate the profit plan to review its impact on profit.) A gross
multiplier would include a factor for most consultants and reimbursable expenses.

A profit plan will provide:

1. Revenue and profit goals to aim for and measure progress
   against
2. Information for pricing new work, including a multiplier, overhead rate, and billing rates
3. Chargeable rates in total and by employee
4. Salary budget in total which provides a target amount for raises
   for the year
5. Overhead and marketing budgets
6. Marketing goals (total volume of work required to achieve goals)

Keep in mind that at year's end it is unlikely that all the targets indicated on your profit plan will be achieved, but as experience grows,
plans will become increasingly more accurate.

	Total ($)	Project expenses ($)	Overhead ($)	Profit ($)
Principal's draw	150,000			
60% project assigned		90,000		
40% unassigned			60,000	
Technical salaries	350,000			
83% project assigned		290,000		
17% unassigned			60,000	
Administrative salaries	100,000			
20% project assigned		20,000		
80% unassigned			80,000	
Total labor	600,000	400,000	200,000	
Non-reimbursables	10,000	10,000		
Overhead	400,000		400,000	
Net profit (before tax)	390,000			390,000
Net fees (revenues)	1,400,000	410,000	600,000	390,000

Income statement

Net fees	$1,400,000
Labor	($400,000)
Non-reimbursables	($10,000)
Gross profit	$990,000
Overhead	($600,000)
Net profit (before tax)	$390,000

Figure 7.4   Sample profit plan and income statement.

Determining a multiplier

Figure 7.4 produces an income statement based upon line 5 of the Profit Plan. This income statement is used to produce percentage targets and a firm multiplier, as shown in Figure 7.5.

Financial ratios for project managers

There are many financial ratios that can be calculated, but several stand out as most important for project managers.

		Target	
		Percent	**Time-card ratio**
Net fees	$1,400,000	100.0	3.500
Labor	($400,000)	28.6	1.000
Non-reimbursables	($10,000)	0.7	0.025
Gross profit	$990,000	70.7	2.475
Overhead	($600,000)	42.9	1.500
Net profit (before tax)	$390,000	27.9	0.975

Figure 7.5   Income statement and percentage targets.

Net profit ratios

This bottom-line analysis of profitability has four variations, each of which will be meaningful to you based upon your method of doing business.

1. *Net profit on net revenues (before distributions).* Net revenues are those earned by the firm after "pass-through" items such as consultants and other reimbursables are deducted. In this ratio, profitability is measured before distributions of profit, bonus, and any other discretionary distributions.

   net profit/net revenues = profit percentage

   This ratio is particularly meaningful for firms that do not regularly pay year-end bonuses or profit sharing and who wish to analyze profit only on their own revenues.

2. *Net profit on net revenues (after distributions).* This ratio is calculated in the same manner as the before-distributions ratio except that bonus, profit sharing, and other discretionary distributions of profit are deducted before the calculation is made. Most firms charge these distributions to overhead. If you pay these various distributions each year and consider them a cost of doing business, then this calculation would be useful.

3. *Net profit on total revenues (before distributions).* For firms that use few outside consultants and have a minimal level of reimbursable expenses, profit measured on total (gross) revenues is more useful:

   net profit/total revenues = profit percentage

4. *Net profit on total revenues (after distributions):* This ratio is calculated in the same manner as ratio 3 above, except that an after-distributions base is used.

Overhead ratios

Overhead evaluation and reduction is one of the most significant concerns for project and firm managers. This important ratio can be calculated two ways:

1. *Overhead rate (before distributions).* As noted earlier, discretionary distributions are customarily charged to overhead. As a result, a before-distributions calculation of overhead will provide an accurate ratio for those firms that do not consider discretionary distributions to be a cost of doing business. The overhead rate is calculated as follows:

    total firm overhead/total firm direct labor
    = overhead rate

    Direct labor is based only on raw labor without fringes (which are charged to overhead).
2. *Overhead rate (after distributions).* This ratio is calculated the same as in ratio 1 except that total firm overhead now includes discretionary distributions of profit.

Net multiplier

The net multiplier is calculated by dividing net revenues by total firm direct labor.

net revenues/total firm direct labor = net multiplier

This ratio indicates the effective markup a firm is achieving for each $1 of direct labor expense (not direct personnel expense, which includes fringe benefits). It is not your target multiplier, but your actual multiplier achieved.

Chargeable rate

The chargeable rate is significant because it measures total (or technical staff) time actually charged to projects (whether billed or not). It measures staff utilization and should be maximized. It is calculated as follows:

total firm direct labor/(direct labor + fringes
+ indirect labor) = chargeable rate

All values in the denominator are based on totals for the firm.

Productivity measures

There are two semi-productivity measures firms could calculate, but recognize that these are imperfect, as noted below.

1. *Net revenues per total staff.* This productivity measure determines how many fee dollars you derive from each staff member:

   net revenue/total staff = net revenues per total staff

   Total staff includes all technical staff, principals, support staff, and all part-timers (adjusted as full-time equivalents).

2. *Net revenues per total staff.* This measure is calculated the same as ratio 1 above, except only technical staff and principals are included in the calculation.

It is actually very difficult to determine the productivity level of design office staff. For years, many CADD vendors and software manufacturers have been promoting the productivity gains achieved by using their equipment and programs. Often, these claims have been based on vague measures. Typically, the time required to produce a detail or sheet of drawings is the yardstick. Unfortunately, these manufacturers fail to measure the time and effort required to achieve the capability to complete the drawing on CADD. They also do not directly relate the economic value of that accomplishment to the firm.

Many firms try to relate productivity to economic value. As discussed above, the most common method of achieving this relationship is to use "net revenues per total staff". This factor relates the net (without consultants and reimbursables) dollar amount of revenues billed (not necessarily collected) per each total staff member.

While this factor ties productivity to economic value, it contains some inherent weaknesses. For example, if the marketing staff within a firm is unable to maintain a steady flow of new work, then it is of little consequence that the technical staff is highly efficient. An additional problem results if project managers or principals are poor negotiators or do not carefully manage the scope of services. As a result, projects may reach maximums more quickly then if project administration was more effective.

A highly productive staff may not translate into profits if project fees are set at an insufficient level to cover the cost structure of the firm. Despite the difficulties in using the revenues-per-employee calculation, it can still be of great value to firm managers. And, it is superior to measures of productivity based simply on time calculations.

Average collection period

This financial measure calculates the length of time required from the date of billing to the date of collection. This measure is directly reflected in cash-flow measures. The two-step calculation is as follows:

Step 1
annual total revenues ÷ 365 (days in the year) = average day's sales

Step 2
average accounts receivable ÷ average day's sales = days

Comparative data

Several industry surveys are available for comparison; see the Bibliography for more information.

Chapter eight

Owner/client concerns

It is crucial for design firms to provide high-quality service to keep their clients satisfied. Many practices obtain the largest part of their work from repeat clients. One financial industry observer has identified five keys to client retention:

1. *Respect.* Treat all clients, whether institutional or individual, with respect. If you respect your clients, they in turn will respect you and your organization. Take the time to find out what they want. Listen to what they have to say. And, give them what they need, not what you assume or think they want.
2. *Communication.* Communication goes hand-in-hand with respect. Keeping everyone well-informed makes the team effort a reality. This includes all aspects of written and oral communication. Timely invoicing, monthly reports, periodic telephone calls, personal visits — everything — must be complete and well prepared. Review each client's status on a regular basis. Contact them and ask for suggestions and comments. Determine if they need anything or if there are problem areas. Communicate both the good and bad news. It is always best that they hear the news from the source.
3. *Service.* Emphasize to your staff that they must offer prompt and complete service to your clients. Research and implement controls and systems to ensure that clients are receiving the best service you can give. This covers everything from telephone procedures to individualized attention from staff. Quick responses to requests and immediate resolutions to problems are critical. Respond to telephone calls in a personalized manner with courtesy and intelligence.
4. *Satisfaction.* A satisfied client translates into loyalty and referrals. The combination of respect, communication, and service encourages a client to remain with you. Satisfaction also means coping

with problems in such a way that all parties achieve their objectives and you maintain your client. Communicate to your staff the necessity of quick resolutions of problems. A problem should never become an issue, at which time it may be too late to avoid damaging your relationship with the client.

5. *Loyalty.* Stand by your clients. Seek ways to offer your help, your resources, and your abilities. Do your best to help them solve their problems and needs within their budget and time frame. Your efforts will be repaid both with client loyalty and with the positive image that is communicated to the entire industry.

Client expectations

It is surprising how many design professionals fail to fully understand their clients' needs and wants. Many designers have little interest in learning about client priorities, methods of operation, or information needs. Some professionals are even condescending to their clients. These designers have the attitude that they know what is best for the client and want no interference. Unfortunately, these designers forget whose project it actually is. Every owner/client is different, and their needs and wants vary. There are, however, some needs that are common.

Clients want a clear point of contact

Owners and designers are often moving in opposite directions on this subject. In recent years, there has been a trend towards encouraging owners to hire all consultants directly. This is due in large measure to liability concerns — designers can be found liable for the errors and omissions of consultants they hire. When a consultant is hired by the client, the prime designer's liability is lessened.

On the other hand, owners are looking for one party to be responsible and in charge of the project. As projects have become increasingly complex, this need has grown. The result has been the development of specialists to fill the gap left by engineers and architects. These project management and construction management firms are now assuming the responsibilities formerly held by many design firms. The result is a deterioration in the designer's role, scope of services, and authority.

A design firm's internal project management structure can also frustrate many owner/clients. As noted previously, owners are looking for one individual to be in charge and responsible for their project. When they have a question or problem, they want to know whom to contact.

Unfortunately, many design firms operate on a crisis management basis. Often, especially in smaller firms, principals try to run projects, bring in new work, and also manage the firm. Rarely do they succeed in all of these activities. A departmentally organized firm can only make the situation worse. In these firms, a series of ever-changing individuals is responsible for the project during its various phases. As a result, there is not a consistent point of contact.

Most owners favor the matrix or strong project manager system of project delivery. In this system, the client knows who is in charge of the project and whom to contact. Nearly all experienced and sophisticated clients endorse this approach, and most employ it in their own organizations.

Clients want to be kept informed

It is common complaint among owners that their design firms fail to keep them adequately informed as to the project's progress and the design options involved. Most feel that the matrix management system gives them the best opportunity to keep current on the project.

In addition to setting up a matrix management system, designers must develop reporting systems, meeting processes, and monitoring tools to keep clients informed. While it is true that some clients are slow to make decisions, it is important to keep them informed. Often, they are slow to make decisions simply because the designer has not kept them informed or neglects a situation until a crisis occurs.

Clients want good cost control

To some owners, designers seem to be unaware of the costs of design decisions. To many designers, some owners want more out of a facility than they are willing to pay for. While this dichotomy may never be resolved, firms that show a concern for construction costs at an early stage of the project are most appreciated by owners.

This situation is particularly acute for public sector clients. Many of these agencies receive their funding based upon an appropriation and find it very difficult to find money to pay for increased construction costs.

Clients want technical competence

Although many designers do not believe it, most owners are more concerned with technical competence and experience than they are with the designer's fee. Clearly, design fees are an issue to many clients;

however, most owners repeatedly indicate that fees are not as important an issue as many designers believe. Many owners look for design firms with quality programs. After all, clients do not need injuries, delays, or lawsuits to hamper the progress of projects, either.

Clients also want to be invoiced regularly

This is essential if they are to plan and manage their own cash flow. Many designers are slow at billing or fail to provide complete and accurate information. The result is a delay in processing invoices and potential irritation to all parties. In a well-managed design firm, it is the project manager who obtains information on billing needs and requirements at the beginning of a project.

What owners should ask designers before selecting them

Several professional societies publish guidelines for owners to use in evaluating and selecting engineers and architects. For example, the AIA publishes *You and Your Architect,* a question-and-answer discussion on getting the most out of the client-architect relationship. Such publications tend to be slanted toward traditional delivery methods (see Appendix A), fail to ask crucial questions, or attempt to encourage the potential client to hire the particular type of professional who prepared the publication.

Every project manager must be prepared to provide a potential or current client with a wide range of information. Failure to provide the requested information completely or honestly may cost you and your firm the contract. Some items a sophisticated client may ask you include:

1. *Experience.* Of particular importance is your firm's experience with similar projects. Do not dwell however, on what you did for others; focus on the problem and potential client at hand. Do not be a firm focused on selling and your own needs; instead, be a marketing-driven firm (focused on the needs of the marketplace).
2. *Pricing justification.* As a project manager, you should fully understand how your firm has arrived at its pricing scheme, including overhead rates, profit targets, overall multipliers, etc. (see Chapter 7). Be prepared to provide hourly rate information (typically pay rates) if requested. If senior management in your firm will not provide you with this information, then you cannot

adequately perform your job as a project manager (see Chapter 10).

3. *Explanation of your project delivery system.* The most successful designer/owner relationships occur when both have similar or complementary systems. For example, you may have a single-discipline, departmentally organized firm, and the client may have a sophisticated project management system. Under these circumstances, it will likely be difficult to work together.

4. *Specific team members.* Some experienced clients require that the project manager and key team members be named in the contract. The design firm is not allowed to change team members without the specific approval of the client. (Some designers wish they could do the same to clients who often change their own project managers or staff.)

5. *Fee.* Although most designers would like to use a qualification-based selection (QBS) process, fee is often a factor. During negotiations, it is vital that a project manager fully understand his firm's pricing scheme, proposed budget (and underlying assumptions), proposed scope of services, change order management process, and any other significant items.

6. *Project management manual.* Experienced clients sometimes request a copy of a design firm's project management manual (see outline in Chapter 5). This manual documents a firm's procedures and can help to reassure a potential client that you have well-thought-out processes, systems, forms, etc. This also gives them the opportunity to understand how your firm operates and how you will work together on the project. Unfortunately, most design firms do not have a written (partial or complete) project management manual. Often, something is simply thrown together when a Request for Proposals (RFP) seeking a copy is received from a potential client. If a firm is considering seeking ISO 9000 certification (see Appendix C), this manual must be prepared.

Commissioning

Commissioning is a concept that has evolved from several sources. First, clients/owners have reduced their own design, construction, facilities, and maintenance staffs, necessitating greater outsourcing. Second, building systems have become more complex, requiring increased expertise to maintain cost-effective operations. Third, clients have become more demanding of their outside design and construction consultants and contractors, expecting increased services.

The American Society of Heating, Refrigerating, and Air-Conditioning Engineers (ASHRAE) has prepared a guideline (1-1996) that defines the concept of commissioning: "Commissioning is the process of ensuring that systems are designed, installed, functionally tested, and capable of being operated and maintained to perform in conformity with the design intent. In this guideline, commissioning begins with planning and includes design, construction, start-up, acceptance, and training and can be applied throughout the life of the building." In essence, it ranges from conception to demolition.

Many design firms, particularly architects, view commissioning as simply another threat to their traditional area of control and responsibility. The truth is, however, that many design firms do not provide this service to their clients. Clearly, the need is there, and if design firms fail to market this service to clients, then other specialists will. Already a group of consultants has appeared called the Commissioning Authority or Commissioning Agents. The commissioning service is typically sold as an additional service at an additional fee. Project managers need to be aware of the opportunities that commissioning provides to both their own firm and to the client.

Partnering

Few industries have such a history of litigation and disputes as that of the construction industry. Many of the contract documents produced by the various professional societies have been drafted specifically to pass risk to others or to spread risk among all parties to the project. Conflict is common; trust is not. Some in the industry have sought to change this to restore trust and a spirit of working together to achieve a common goal. In the past few years, this has become embodied in the partnering concept.

Tom Warne, current director of the Utah Department of Transportation, prepared a manual on partnering for the American Society of Civil Engineers (ASCE) called *Partnering for Success* (see the Bibliography for more information). In this manual, he wrote, "Partnering recognizes that there are many stakeholders on any given construction project. A stakeholder may be an owner, a prime contractor, a design engineer or architect, a subcontractor, a supplier, a local community or business group, a governmental agency, or any one of a host of others."

A key to the partnering process is for all of the parties involved in a construction project to come together and examine their goals and objectives to determine common ground. Typically, these goals and objectives include a desire for quality, profitability, safety, on-time completion, and other similar items. The partnering team then determines how to work together to achieve these goals.

This is done initially through a partnering conference where all of the key players, including their respective project managers, meet for 2 or more days. Each organization represented presents their goals, objectives, concerns, methods of operation, etc. Common ground is identified and agreement is reached on dispute and issue resolution. This is embodied in a Partnering Charter that all sign and agree to follow. The initial partnering conference is best led by an outside facilitator who is objective and can direct the process. It should never be led by the client. As part of this conference, the parties agree to establish a regular partnering meeting process throughout the life of the project.

In order to make partnering a success, several additional items should be considered.

1. Partnering requires the strong backing of the client. If they believe in it and support it, then it will succeed.
2. The client has to be willing to pay for the partnering process by hiring a facilitator and by paying for the time of the team members for participating.
3. The design and construction team members must support partnering from the top down. If senior management does not provide support, then the process becomes much more difficult.
4. Project managers must be regular participants in partnering meetings.

Role of specialists hired by owners/clients

A wide range of specialists has appeared on the construction scene in recent years. Most are hired by the client to aid in cost control and/or provide management skill lacking in the owner's organization. How successful these specialists are is a subject of great debate.

Project management consultants are firms that specialize in managing a particular project for a client who lacks the skill, staff, time, or interest to do so. For example, a hospital adding a doctors' office building might turn to a project management consultant to help define the scope, determine funding needs and devices, hire architects and engineers, etc.

Program management consultants are similar to project management consultants except that they manage an entire building program. For example, a school district renovating many schools may hire a program management consultant.

Design builders are used by owners seeking "one-stop shopping" for both the design and construction. The belief is that communication and cost control are improved when both functions are within the same organization. A corporation requiring a new manufacturing facility may find this a very effective approach.

Construction managers typically justify their fee by the cost savings they can identify during the design and construction process. This may occur based on a product substitution, alternative erection process, Value Engineering (see Appendix B), or other methods. Construction managers can be very effective for large, complex projects such as a significant public building, highway reconstruction, or other similar jobs.

Chapter nine

Scope determination

There are many methods that firms use to determine the scope of services on a project. Many work with detailed checklists. Some use approximate guidelines, and others simply use the broad definition of services offered in AIA and other organizations' standard form contracts.

A preferred method requires detailed analysis of the activities required to complete the project. This list of activities can be reviewed with the client and a determination made as to those for which the designer is responsible. An excellent planning tool for determining the scope of services is the AIA's Compensation Guidelines System (refer to AIA standard contract form B163). This system provides a series of phase-by-phase formats listing detailed activities. Although architectural descriptions are shown, minor changes could easily make the formats useful for engineers or interior designers. Figure 9.1 shows a sample worksheet that can be used for determining the project scope.

Typically, the architect uses this series of forms to inform the client of the variety of available services. Additional services such as "graphic design" could be added based upon either client request or suggestion by the designer. Disagreements over providing a service or responsibility for an activity can be minimized by notation under the appropriate heading across the top. Separate compensation methods could be indicated as shown, although this should be kept reasonably consistent to avoid confusion.

Although this is an excellent tool for determining a scope of services, this system has significant drawbacks as a management and monitoring tool. Its complexity makes filling out time sheets in such detail nearly impossible. Project status reports showing this extensive detail would be impossibly complex.

SCHEDULE OF DESIGNATED SERVICES WORKSHEET	By Architect	By Architect, as Outside Services		By Owner, Coordinated by Architect	By Owner		By Architect, as Additional Service*	Not to be Provided	Method of Compensation
PHASE 5: CONSTRUCTION DOCUMENTS SERVICES									
Project___ Project #___ Date___ Owner___ Architect___									
5.01 Project Administration									
5.02 Disciplines Coordination/Document Checking									
5.03 Agency Consulting/Review/Approval									
5.04 Owner-supplied Data Coordination									
5.21 Architectural Design/Documentation									
5.22 Structural Design/Documentation									
5.23 Mechanical Design/Documentation									
5.24 Electrical Design/Documentation									
5.25 Civil Design/Documentation									
5.26 Landscape Design/Documentation									
5.27 Interior Design/Documentation									
5.28 Materials Research/Specifications									
5.30 Special Bidding Documents/Scheduling									
5.32 Statement of Probable Construction Cost									
5.33 Presentations									

METHODS OF COMPENSATION

A = Multiple of Direct Salary Expense   E = Stipulated Sum
B = Multiple of Direct Personnel Expense   F = Hourly Billing Rates
C = Professional Fee Plus Expenses   G = Multiple of Amounts Billed to Architect
D = Percentage of Construction Cost   H = Other:

*Requires separate authorization and compensation

AIA° FORM F815 · DESIGNATED SERVICES WORKSHEET, PHASE 5 · JANUARY 1978 EDITION · © 1978   **THE AMERICAN INSTITUTE OF ARCHITECTS**
THE AMERICAN INSTITUTE OF ARCHITECTS, 1735 NEW YORK AVE., N.W., WASHINGTON, DC 20006   **FINANCIAL MANAGEMENT SYSTEM**

Figure 9.1   Sample completed schedule of designated services worksheet (form F818). (From *Managing Architectural Projects: The Process,* The American Institute of Architects, Washington, D.C., 1984. With permission.)

Dividing contracts

Many design firms are dividing their contracts into three or more separate parts — for example, scope determination, programming, and possibly preliminary design work are handled as one contract and are often billed to the client on a time-card basis. A second contract is then prepared once the detailed requirements of the project have been determined. A third contract is prepared for the construction administration phase. Some clients even hire outside specialty firms for this last activity.

Selecting outside consultants

The early selection of outside consultants to join the project team is an extremely important decision. This allows their input into the scope definition process and ensures that all parties are in agreement on responsibilities and financial issues. Many designers work with consultants with whom they have a long-standing relationship. Unfortunately, these individuals or firms may not have the best technical qualifications or be best suited to meet the clients' needs.

The lead design firm must be very thorough and careful in the selection of outside consultants to join the project team. Failure to exercise proper care could result in delays, lower profits, or even lawsuits.

The selection process

Every design firm that recommends consultants to clients should have a well-established process to evaluate these firms; otherwise, you may be doing a disservice to your client and to yourself. This evaluation process should include the following steps:

1. *Establish a resource file of capable consultants.* This should include information on their specialty, if any, as well as any other pertinent data. This material should be updated regularly by your marketing staff or principals. This file is much like that maintained by many of your clients on you and your peers.
2. *Develop an evaluation system for consultants who have worked on your past projects.* This will provide ready reference material on their performance, staffs, methods of operation, etc.
3. *Maintain materials and an evaluation on consultants with whom you are very familiar.* You may, however, include them on your short list without an extended evaluation. Always consider other con-

sultants. You may find them better able to meet your (and your clients') needs.

4. *Collect information on consultants with whom you have not worked.* This includes:

    a. A list of references of several other prime design firms with whom they have worked.

    b. Biographical data on individuals who will be assigned to your project — obviously, this must include information that convinces you of their managerial or technical capability.

    c. Information on the consultant's financial ability — if possible, obtain a financial statement. Your primary concerns are that the consultant can afford to add staff if necessary and can provide the required project manpower. It is also important that the consultant can afford additional office space, if needed, and can meet the higher professional liability premiums that the additional work may require.

    d. Data on the consultant's internal management structure and project delivery methods — it is a great help if their method of operation is similar to yours. For example, your project manager can be more effective if he or she has a clearly designated counterpart in the consultant's office.

    e. Your method of operation for the potential consultant — if you have a project management manual, provide a copy for

---

Successful negotiating

A successful project manager is a successful negotiator; some suggested negotiating concepts follow:

1. Know your costs. It is difficult to successfully negotiate if you do not know your own cost structure.

2. Have a floor or walk away price. Anytime you want something too badly, you have significantly weakened your position.

3. Know the proposed scope of services. It is impossible to negotiate unless you know what you are proposing to do.

4. Try to negotiate on your own turf. This allows you to control the flow of the negotiation and have needed resources at your fingertips.

5. Have staying power. Stay with it until an acceptable conclusion is reached or you realize you need to walk away. Do not tip your hand if you are in a time crunch.

them as a reference guide to your operations. Require the return of this manual in the event that the consultant is not selected.

f. Information on the consultant's specific computer software and hardware and other technical capabilities — it is important to have compatible software and systems. This is especially true if CADD systems are to be used to produce the work.

g. Certificate of insurance for each potential candidate, indicating the existence of professional liability coverage and the amount. Failure to do so may leave you as the "deep pocket" in the event of a lawsuit.

h. A summary of the potential consultant's quality management/assurance procedures will minimize client problems, contractor disagreements, and potential litigation.

5. *Prepare a clear and complete contract.* Never operate on a handshake or verbal agreement. This is especially true on change orders. Without effective change-order management, you may find your consultants undertaking work well beyond or not in accord with your client's wishes. Also, without a firm agreement on fees for changes, you may find the consultant invoicing you for an amount in excess of that which you will receive from the client.

6. Learn about your opponent. Know their needs, weaknesses, etc. Recognize that they are the opponent until an agreement is reached.

7. Watch out for "nibbles". These are the little changes (that add up) that occur after an agreement is reached.

8. Knowledge is power; get as much as you can, and give away as little as possible.

9. Be reasonable. If you perceive an offer to be fair then accept it — trying to get the last nickel can be counterproductive and antagonistic to people you will have to work with later.

10. Always put "padding" in your time, scope, and cost proposals to allow concessions and movement.

11. When you make a concession, make sure your opponent knows you are making it.

12. Be sure to declare your opponent the winner!

Your agreement with outside consultants must be very specific as to billing and payment terms and conditions. Many prime design firms are extremely slow to pay their consultants. The prime must be prepared to pay his consultants immediately after receiving payment from the client (subject to severe penalty if not done). This is a particularly upsetting subject to many engineers who work with architects. Some architects use money that rightfully should be paid to their engineers to finance their own poor management.

Chapter ten

Project budgeting

The preparation of budgets and the monitoring of costs are major functions of project management. Although they are often performed separately, they are basic to a firm's profitability and success in providing effective professional services.

Budgeting

The project budgeting process has two interdependent aspects: (1) estimating compensation before negotiating with the client, and (2) determining the actual detailed project budget by project phase or activity. Neither can be adequately performed without the other.

Experience indicates that too many design firms estimate compensation on a percentage of estimated construction cost basis, negotiate that fee with the client, and then attempt to fit their actual costs into that fee maximum. There are, of course, many other methods of job costing. These include time-based methods such as straight time card and time card to a maximum. Others are task based, such as fees based on average labor cost per sheet of drawings, while still others may not have any real basis, such as those using lump-sum or a flat fee.

The preferred method is to determine a firm's actual costs first and then communicate the proposed fee to the client in whatever form required (percentage or construction cost, lump-sum, cost-plus, multiplier, or hourly rate). For example, Engineer Smith is competing against Engineer Jones for a commission. The prospect has retained engineers in the past and has always negotiated a percentage of construction cost contract. Smith cannot present the prospect with an elaborate method of determining fair compensation if Jones simply offers a 5% fee to perform the work.

Smith's approach should be to determine actual costs while recording areas in which he has room to adjust his fee (perhaps by altering the

scope of services he will provide), then translate this fee into an estimated percentage of construction cost by the following formula:

$$\text{cost (including profit)} =$$
% of construction estimated construction costs (fee)

As mentioned, Jones has already set his fee at 5%. Suppose Smith's
analysis results in a fee of 6%. His options are

1. Alter the scope of services provided.
2. Go with the 6% fee.
3. Determine that his time is better spent seeking other projects that
   will provide the fee he desires.
4. Take the project in the hope that he can complete it within the
   budget (without damaging quality and service).
5. Remind the client that a lower fee may mean lower service.

The decision is his. The key is knowing his costs.

Where do firms obtain the information to determine these actual
costs? There are many sources including:

1. The firm's experience and records on previous projects of similar
   scope (see Figure 10.1 for the suggested content of this historical
   file).
2. Cost per sheet (sheet counts), but only if this information is
   based upon documented cost per sheet for services by project
   type. To use the cost per sheet from an old firm or a neighbor's
   is not adequate. Cost must relate to the firm's level of productivity, projects, etc.
3. A manager's knowledge of similar projects. This is often the
   single best source of data if combined with recorded historical
   information. In many firms, the project manager's private files
   are often more complete than the firm's central files. With staff
   turnover, however, much of this information is often lost.
4. Cost-based compensation guidelines (see Chapter 9). This system, developed by the AIA, provides an extensive breakdown
   of all possible services provided by a design firm. It is an excellent method for isolating the various activities you must perform and for determining a cost for each activity. Many firms
   have successfully used this system as an educational tool to
   show their clients the multitude of tasks that are required to
   meet the client's needs. Figure 10.2 shows a sample budgeting
   form from this system. This form integrates with AIA contract

---

Project data
Project name
Project address
Related project references
Referenced contracts and addresses
Owners name and address
Gross square footage
Construction costs (budgeted, contracted, final)
Cost/gross square footage (exclusive of sitework)
Cost analysis (if available)
Program summarization (major areas)
Fee (budgeted and final)
Project starting date
Construction starting date
Final completion date
Final project summary report
Consultants (including addresses and telephone numbers)
Project manager's statement (one page; should cover the significance
        of the project, personnel involved, contractor selection process,
        scope of services, awards, information of pubic interest, potential
        of owner recommendation, etc.)
Project designer's statement (brief statement of the program, design
        constraints, objectives, materials, etc.)
Copy of the last Certificate of Payment, including list of contractors
List of subcontractors and material suppliers (should include product
        manufacturers for publication and field rep's evaluation of each
        subcontractor and supplier in regard to delivery time, quality of
        workmanship, recommendations for use on other projects, etc.)
Site plan (8-1/2 × 11) showing building configuration
One 8 × 10 photograph

Figure 10.1 Information required for a supplementary completed project file (a single folder containing project information is required for each current job and each recently completed project).

document number B163. It is available from local AIA bookstores or from the AIA in Washington, D.C. (202-626-7300). Engineers and others can easily use the form by simply changing the descriptions.

5. Percentage of construction costs, but only if it is based on the firm's actual recorded experience with the building type, with similar clients, in similar locations, etc. This method may expose

PHASE COMPENSATION WORKSHEET

PHASE 5: CONSTRUCTION DOCUMENTS SERVICES

Project _____
Project # _____ Date _____
Owner _____
Architect _____

SERVICE columns:
5.01 Project Administration
5.02 Disciplines Coord./ Document Checking
5.03 Agency Consulting/ Review/Approval
5.04 Owner-supplied Data Coordination
5.21 Architectural Design/ Documentation
5.22 Structural Design/ Documentation
5.23 Mechanical Design/ Documentation
5.24 Electrical Design/ Documentation
5.25 Civil Design/ Documentation
5.26 Landscape Design/ Documentation
5.27 Interior Design/ Documentation
5.28 Materials Research/ Specifications
5.30 Special Bidding Documents/Scheduling
5.32 Statement of Probable Construction Cost
5.33 Presentations

IN-HOUSE PERSONNEL (Hrs./$)
SUB TOT.
OUT-SIDE (Hrs./$)
TOT. (Hrs./$)

TOTAL HOURS | TOTAL DOLLARS | ITEM | LINE

ITEM	LINE
@ $	1
@ $	2
@ $	3
@ $	4
@ $	5
Direct In-house Salary Expense	6
Direct Personnel Expense	7
Indirect Expense	8
Other Nonreimbursable Direct Expense	9
Total In-house Expense	10
Outside Services Expense	11
Estimated Total Expense	12
Contingency	13
Profit	14
Proposed Compensation	15
Estimated Reimbursable Expense	16

REMARKS

AIA FORM F825 · PHASE COMPENSATION WORKSHEET, PHASE 5 · JANUARY 1978 EDITION · © 1978
THE AMERICAN INSTITUTE OF ARCHITECTS, 1735 NEW YORK AVE., N.W., WASHINGTON, DC 20006

AIA THE AMERICAN INSTITUTE OF ARCHITECTS
FMS FINANCIAL MANAGEMENT SYSTEM

Figure 10.2 Sample completed phase compensation worksheet (form F825). (From *Managing Architectural Projects: The Process,* The American Institute of Architects, Washington, D.C., 1984. With permission.)

the firm to excessive loss if costs run higher than the fee or if the construction cost is reduced (either by redesign or low bids). At the same time, you should not try to make a windfall profit on a higher than anticipated construction cost. Your job is to reduce costs not increase them to enhance your fee.

Several other considerations are important in developing project budgets. Most designers neglect to forward price their services. With inflation, the cost of labor and overhead will increase over the life of a project. Failure to anticipate these increases can easily turn a profit into a loss. It is important to determine the level of service the firm can provide based upon a client's needs and ability to pay. Clients with a greater ability to pay can receive more elaborate levels of service. Not all clients need to receive the same level of service. The firm has to maintain quality and service but must know when the fee will not support even minimum standards and must be prepared not to seek the commission. To do otherwise is a disservice to the firm and to the client.

Project cost plan

A surprising number of firms do not have a formal method of developing a project budget. Often, for small projects, a detailed breakdown may not be necessary, but the basic cost areas should be considered. All larger projects (even open-ended, time-card, or multiplier projects where an obligation to the client to control costs exists) should be controlled by a formal budget. The more successful firms often incorporate techniques to provide cross-checks using several budgeting methods.

Figure 10.3 shows a project cost plan form incorporating a percent of construction cost cross-check (estimated construction cost). Other major sections include in-house services, direct (non-reimbursable) costs that come out of fees, profit, and reimbursable costs. This form is blank to permit copying for immediate use. Figures 10.4 through 10.8 trace the development of a hypothetical project budget.

Using detailed worksheets, client input, previous experience, and other methods, a general estimate of construction costs by major area is developed. This estimate is then used to determine an estimated fee based upon the firm's historical percentage of construction cost database. The estimate is used only as a cross-check with other methods of estimating compensation.

Services provided within the firm are determined by a careful analysis of the scope of the project (Figure 10.5; the breakdown of services shown is for illustration only). Hours to complete each task are

Date _____   **Project no.** _____

Scope of project _____

Project name _____

Estimated construction cost (sample breakdown only)

Site work	$_____	Electrical	$_____
General	$_____	HVAC	$_____
Plumbing	$_____	Kitchen	$_____
Elevator	$_____	Other	$_____
Total	$_____		

Services provided within the firm (sample breakdown only)

Pre-design	_____ hrs @ $_____/hr = $_____
Site analysis	_____ hrs @ $_____/hr = $_____
Schematic design	_____ hrs @ $_____/hr = $_____
Design development	_____ hrs @ $_____/hr = $_____
Construction documents	_____ hrs @ $_____/hr = $_____
Bid/negotiation	_____ hrs @ $_____/hr = $_____
Construction administration	_____ hrs @ $_____/hr = $_____
Post-construction	_____ hrs @ $_____/hr = $_____
Supplementary services	_____ hrs @ $_____/hr = $_____
Total labor	$_____
Overhead (factor _____%)	$_____
Total labor and overhead	$_____

Non-reimbursable direct costs (costs that come out of fee)

Consultant #1	$_____
Consultant #2	$_____
Consultant #3	$_____
Consultant #4	$_____
Other consultants	$_____
Subtotal consultants	$_____
Travel	$_____
Reproduction and supplies	$_____
Models and photographs	$_____
Telephone and telegraph	$_____
Other	$_____
Subtotal non-reimbursables	$_____
Total non-reimbursable direct costs	$_____

Profit

Total labor, overhead, and direct	$_____
Contingency	$_____
Profit	$_____
Total fee for services	$_____
Markup on reimbursables	$_____
Fee quoted client	$_____

Reimbursables

Total (budgeted or by contract)	$_____
Total compensation	$_____

Comments and notes _____

_____

Figure 10.3   Project cost plan.

Chapter ten: *Project budgeting*

Date ___November 30, 1999___   **Project no.** _____02799_____

Scope of project _Provide complete design for City Hall alterations_____

Project name _____City Hall alterations; Any City, Texas_____

Estimated construction cost

Site work	$25,000	Electrical	$10,000
General	$80,000	HVAC	$40,000
Plumbing	$10,000	Kitchen	$5000
Elevator	$20,000	Other	$10,000
Total	$200,000		

Figure 10.4 Project cost plan (part one).

estimated and multiplied by the average raw labor rate (not direct personnel expense factor — labor plus fringes) of all individuals who will be working on that phase. Each of these line items is totaled to obtain the total labor amount of $2612 shown.

The firm's overhead is determined by multiplying labor by an overhead factor (in this example it is 150%). Hence, the overhead factor line is calculated by the following:

$$\$2612 \times 150\% \ (1.5) = \$3918$$

Date _____November 30, 1999_____   **Project no.** _____02799_____

Scope of project ____Provide complete design for City Hall alterations_____

Project name _____City Hall alterations; Any City, Texas_____

Estimated construction cost

Site work	$25,000	Electrical	$10,000
General	$80,000	HVAC	$40,000
Plumbing	$10,000	Kitchen	$5000
Elevator	$20,000	Other	$10,000
Total	$200,000		

Services provided within the firm (sample breakdown only)

Pre-design	20 hrs @ $10/hr = $200
Site analysis	24 hrs @ $8/hr = $192
Schematic design	30 hrs @ $10/hr = $300
Design development	40 hrs @ $7/hr = $280
Construction documents	160 hrs @ $5/hr = $800
Bid/negotiation	12 hrs @ $12/hr = $120
Construction administration	90 hrs @ $8/hr = $720
Post-construction	_____ hrs @ $_____/hr = $_____
Supplementary services	_____ hrs @ $_____/hr = $_____
Total labor	$2612
Overhead factor (150%)	$3918
Total labor and overhead	$6530

Figure 10.5 Project cost plan (part two).

Total labor and overhead is determined by adding $2612 and $3918 to equal $6530.

Figure 10.6 adds project-related non-reimbursable direct costs that come out of the fee. The consultant cost budget should be determined by written agreement with each consultant. Other non-reimbursable direct costs should be determined by a careful analysis of expected costs. All costs in this section should be kept to a minimum, as those costs reduce potential profit for the firm. As shown in Figure 10.6, non-reimbursable direct costs total $2600. Too often, profit is considered the amount left over when the project is complete. Profit should be planned for, as shown in Figure 10.7. The first step is to determine costs by adding together total labor, overhead, and non-reimbursable direct costs:

Direct personnel expense

Unique to architectural firms is the use of the direct personnel expense (DPE) factor. DPE bases the calculation of the hourly job cost rate for an employee on raw labor plus all statutory and discretionary fringe benefits. Statutory fringes such as state and federal unemployment insurance taxes, the portion of social security paid by the firm, and workmen's compensation insurance are paid by virtually all design firms. The level of these, however, varies greatly by location. Discretionary fringes include major medical insurance, long-term disability insurance, dental insurance, life insurance, vacation, sick leave, holiday pay, and a great variety of other benefits. Not only do the components of the DPE factor vary, but the level of each also varies from firm to firm, among employees of the same firm, and from period to period.

The use of DPE factors has long been incorporated in standard American Institute of Architects contract forms, including the 1997 revision to the B141. Typically, the DPE factor is multiplied by a factor that covers non-fringe benefit overhead expenses, profit, and non-reimbursable (direct) consultant and non-consultant expenses. Most other design professionals simply multiply raw labor by a factor that includes all overhead costs including fringes, profit, and non-reimbursable (direct) expenses. Raw labor is defined as the dollar amount job-costed to a project even if it is different than the employee's actual pay rate.

In many cases, a firm using a DPE factor as a base will have what appears to be a lower multiplier. This may be of great value to a firm in a price competitive situation trying to provide or give the illusion of a lower multiplier. (The use of a multiplier is particularly significant in time-card or time- card to a maximum contracts).

This edge, however, may only be an illusion. It is very possible that a firm using a DPE base may actually have a higher multiplier than a firm using raw labor. For

$$\text{total labor} = \$2612$$

$$\text{overhead factor} = \$3918$$

$$\text{direct costs} = \$2600$$

$$\text{total} = \$9130$$

To this total add factors for contingencies (3% is a generally accepted standard) and profit (17 to 20% is generally targeted). Hence,

$$3\% \times \$9130 = \$274; \; 17\% \times \$9130 = \$1552$$

Thus, the total fee for services is

$$\$9130 + \$274 + \$1552 = \$10,956$$

---

example, Firm A has a DPE factor of 1.4 (if raw labor is a dollar, then fringes are another 40 cents) and a multiplier for non-reimbursables, profit, and overhead of 2.5 times DPE. Firm B has a raw labor rate of 1.0 and a multiplier of 3.2 times raw labor. Firm A appears to have the lower multiplier (2.5 vs. 3.2), but on closer examination, Firm A's multiplier is actually 3.5 (1.4 times 2.5). To the unaware client, Firm A appears to be less expensive, while the reverse is actually true.

In a competitive situation, it is the wise negotiator who presents his firm in the best possible light. Translating and presenting a cost structure to a potential client using DPE could be to the firm's advantage. Significant problems can be created, however, if the firm attempts to use DPE-based data for internal management and permanent records. With the fringe benefit mix varying from employee to employee, project to project, and period to period, the firm's management is never certain (without extensive research) of the actual base labor dollars required and the actual cost of doing the project. DPE-based records are nearly useless for building vitally important historical databases, as even averages are rendered meaningless by a changing package of fringe benefits. By removing fringes from overhead and including them in the DPE factor charged (usually this is done by calculating a firm-wide fringe factor to be added to raw labor), the actual overhead cost associated with a project can be distorted.

Some firms avoid this distortion by providing each employee with his or her own specific fringe factor. Extensive research would still be required, however, to make use of the firm's data base. The best solution is to keep all records on a raw labor base and provide clients with an equivalent DPE multiplier only when competitive conditions make it necessary.

Date _____ November 30, 1999 _____ Project no. _____ 02799 _____

Scope of project ____ Provide complete design for City Hall alterations ____

Project name _____ City Hall alterations; Any City, Texas _____

Estimated construction cost

Site work	$25,000	Electrical	$10,000
General	$80,000	HVAC	$40,000
Plumbing	$10,000	Kitchen	$5000
Elevator	$20,000	Other	$10,000
Total	$200,000		

Services provided within the firm (sample breakdown only)

Pre-design	20 hrs @ $10/hr = $200
Site analysis	24 hrs @ $8/hr = $192
Schematic design	30 hrs @ $10/hr = $300
Design development	40 hrs @ $7/hr = $280
Construction documents	160 hrs @ $5/hr = $800
Bid/negotiation	12 hrs @ $12/hr = $120
Construction administration	90 hrs @ $8/hr = $720
Post-construction	_____ hrs @ $_____/hr = $_____
Supplementary services	_____ hrs @ $_____/hr = $_____
Total labor	$2612
Overhead factor (150%)	$3918
Total labor and overhead	$6530

Non-reimbursable direct costs (costs that come out of fee)

Consultant #1	$800
Consultant #2	$600
Consultant #3	$600
Consultant #4	$_____
Other consultants	$400
Subtotal consultants	$2400
Travel	$_____
Reproduction and supplies	$100
Models and photographs	$50
Telephone and telegraph	$30
Other	$20
Subtotal non-reimbursables	$200
Total non-reimbursable direct costs	$2600

Figure 10.6 Project cost plan (part three).

Note that if the calculation for profit is made by first determining a total fee for services (perhaps on a percentage of construction cost) and then taking 17% for profit (off the top), the dollar amount for profit would be much different:

$$\$10,956 \times 17\% = \$1863$$

Compare this amount with the $1552 determined by the cost-based method ($311 less). The result is a higher profit budget, but a lower

Date _____ November 30, 1999   **Project no.** _____ 02799 _____

Scope of project _____ Provide complete design for City Hall alterations _____

Project name _____ City Hall alterations; Any City, Texas _____

Estimated construction cost (sample breakdown only)

Site work	$25,000	Electrical	$10,000
General	$80,000	HVAC	$40,000
Plumbing	$10,000	Kitchen	$5000
Elevator	$20,000	Other	$10,000
Total	$200,000		

Services provided within the firm (sample breakdown only)

Pre-design	20 hrs @ $10/hr = $200
Site analysis	24 hrs @ $8/hr = $192
Schematic design	30 hrs @ $10/hr = $300
Design development	40 hrs @ $7/hr = $280
Construction documents	160 hrs @ $5/hr = $800
Bid/negotiation	12 hrs @ $12/hr = $120
Construction administration	90 hrs @ $8/hr = $720
Post-construction	_____ hrs @ $_____/hr = $_____
Supplementary services	_____ hrs @ $_____/hr = $_____
Total labor	$2612
Overhead factor (150%)	$3918
Total labor and overhead	$6530

Non-reimbursable direct costs (costs that come out of fee)

Consultant #1	$800
Consultant #2	$600
Consultant #3	$600
Consultant #4	$_____
Other consultants	$400
Subtotal consultants	$2400
Travel	$_____
Reproduction and supplies	$100
Models and photographs	$50
Telephone and telegraph	$30
Other	$20
Subtotal non-reimbursables	$200
Total non-reimbursable direct costs	$2600

Profit

Total labor, overhead, and direct (80%)	$9130
Contingency (3%)	$274
Profit (17%)	$1552
Total fee for services (100%)	$10,956
Markup on reimbursables	$544
Fee quoted client	$11,500

Figure 10.7 Project cost plan (part four).

amount available to budget for other items. But this is trying to fit a square peg in a round hole. The cost-based method is preferred, because it determines profit based on the actual cost of doing the work and not on an artificially remaining amount after profit is deducted.

The final items to consider are reimbursable expenses over and above the basic fee for service. These expenses are developed from a separate schedule. A percentage markup (often ranging from 5 to 25%) sometimes is added to this amount (see sidebar).

In Figure 10.8, total reimbursable expenses are $3000 and the markup on these reimbursables is $544 or about 18%. Reimbursables include many of the same types of items (including consultants) shown under direct costs. Hence, a line item for a soils consultant may be found under both the direct and reimbursable sections, as well (the same is true for other items), depending upon the contract with the client.

This completes the basics of project budgeting. After determining the actual costs, a fee may be presented to the client in whatever form he or she desires. To adjust the fee, modify the relevant section of the project cost plan and revise the subsequent figures.

Revealing salary information to project managers

Some design firm principals are reluctant to provide salary data to project managers. This is done in the mistaken belief that the confidentiality of pay rates is being protected. Unfortunately, this places an unfair burden on project managers. They are expected to manage costs on projects without knowing what these costs are.

You cannot manage a project by using hours alone. It is extremely likely that meeting the hours budget will result in a dollar amount at great variance with the budget. This occurs because the mix of individuals actually working on a project and their pay rates is often at great variance with those in mind when a fee is negotiated. Managing on hours fails to provide the project manager any frame of reference for controlling project costs.

Some design firms have pay rates that differ from the employees' actual job cost rate. Standardized employee classifications may exist such as Technical I, Technical II, etc. where all workers in a class are job costed at the same rate.

In other firms, job cost rates are rounded to the nearest dollar or 10 dollars to promote standardization or to ease budgeting and management. Both result in a variance from pay rates. In these cases, the PM does not need to know pay rates, but only job cost rates. However, senior firm managers must recognize that this variance will make it more difficult for project managers to control costs effectively.

Date _____ November 30, 1999 _____ **Project no.** _____ 02799 _____

Scope of project _____ Provide complete design for City Hall alterations _____

Project name _____ City Hall alterations; Any City, Texas _____

Estimated construction cost (sample breakdown only)

Site work	$25,000	Electrical	$10,000
General	$80,000	HVAC	$40,000
Plumbing	$10,000	Kitchen	$5000
Elevator	$20,000	Other	$10,000
Total	$200,000		

Services provided within the firm (sample breakdown only)

Pre-design	20 hrs @ $10/hr = $200
Site analysis	24 hrs @ $8/hr = $192
Schematic design	30 hrs @ $10/hr = $300
Design development	40 hrs @ $7/hr = $280
Construction documents	160 hrs @ $5/hr = $800
Bid/negotiation	12 hrs @ $12/hr = $120
Construction administration	90 hrs @ $8/hr = $720
Post-construction	_____ hrs @ $_____/hr = $_____
Supplementary services	_____ hrs @ $_____/hr = $_____
Total labor	$2612
Overhead factor (150%)	$3918
Total labor and overhead	$6530

Non-reimbursable direct costs (costs that come out of fee)

Consultant #1	$800
Consultant #2	$600
Consultant #3	$600
Consultant #4	$_____
Other consultants	$400
Subtotal consultants	$2400
Travel	$_____
Reproduction and supplies	$100
Models and photographs	$50
Telephone and telegraph	$30
Other	$20
Subtotal non-reimbursables	$200
Total non-reimbursable direct costs	$2600

Profit

Total labor, overhead, and direct (80%)	$9130
Contingency (3%)	$274
Profit (17%)	$1552
Total fee for services (100%)	$10,956
Markup on reimbursables	$544
Fee quoted client	$11,500

Reimbursables

Total (budgeted or by contract)	$3000
Total compensation	$14,500

Comments and notes _____

_____

Figure 10.8 Project cost plan (part five).

Value pricing

Value pricing and value marketing are certainly not new concepts. The architectural profession in particular has been an advocate of value pricing, as fees have declined and competition from non-traditional providers of design and construction services has grown. Obviously, everyone wants to be valued for their knowledge and expertise. The

---

Reimbursable markups

Many design firms mark up reimbursables. The typical markup is 10%, with a range from 5 to 25%. Often, firms attempt to justify this markup by claiming that they are incurring additional administrative, clerical, and coordination time (direct labor time by technical staff and project managers) and expense in processing these items. This claim is often difficult to substantiate, since these costs are normally charged to overhead. If they are being charged to clients as a markup, then the overhead rate charged should be reduced accordingly. Failure to do so will result in double charging clients for the same item. If the time or expenses being incurred in processing these items is extreme, they should be charged as a direct project cost and not as a markup or overhead.

There are, however, situations where a markup is justified. A firm's opportunity cost of money tied up in paying for reimbursables prior to payment by the client is a justified charge. For example, if a firm could earn 6% interest on money in an investment account, but instead must use it to pay reimbursable vendor bills or jeopardize its own credit standing, there is an opportunity cost incurred.

A firm also incurs a real cost when it must borrow money to pay for reimbursable vendor bills for a slow-paying client. When clients are willing to provide a retainer or other guarantee of prompt payment, a markup can be eliminated.

Lastly, an argument could be made for charging a markup on an item to avoid raising the firm's cost structure and penalizing all of your clients. For example, a particularly large project may significantly increase your professional liability premium. This may result in raising your overhead rate, thus making you more expensive to all of your current and potential clients. By directly charging the responsible client for this cost, perhaps by treating it as a reimbursable item with a markup, you avoid this problem.

In cases where clients, in the negotiation process, refuse to allow you a sufficient overhead rate to cover actual costs, you certainly are justified in seeking a markup to cover the clearly identifiable costs of processing reimbursables. It is most important, however, that the firm's negotiators completely understand their cost structure. Markups should not be established simply because competitors are doing so or because an unaware clients permits it. Markups must be justified to be valid.

crucial issue is perception of value. Architects and engineers naturally value their skills and knowledge highly. And, just as obviously, the marketplace is discounting their value or purchasing from competitors such as design builders, project management consultants, program managers, construction managers, etc.

In mid-1997, the AIA published a survey on services offered by architects. It is instructive as to how architects (and likely engineers) are addressing this issue. In comparing services offered in 1990 and in 1996, the survey found a decline from 78% to 60% in architectural firm billings generated by offering design services, while construction-related billings rose from 5% in 1990 to 18% in 1996. Clearly, the marketplace is forcing design firms not only to undertake value marketing, but also to provide real services needed by clients/owners. Project managers are a crucial link in this process by serving clients successfully, participating in the marketing process, and by being alert to new or additional services needed by their current clients.

Higher fees or fees based upon a designer's perceived self-value cannot be achieved without value marketing. Clients are sophisticated, and project managers must prove their value and that of their firm every day.

Chapter eleven

Project scheduling

Project schedules are important and highly useful tools for designers. The preparation of schedules allows for detailed planning of work activities and provides a device for communicating critical dates and activities to clients, consultants, and the internal project team. In addition, schedules provide a tool to help program the project, test alternative approaches, and evaluate job performance.

The development of a project schedule is an aid — not a substitute — for project management. It is a useful method for monitoring percent complete, offers guideposts for the project team, and can show the real possibility of meeting deadlines.

The first step in developing a project schedule begins with defining the planning units. This may be based on geography (locations), function (architect, engineer, etc.), or phase (design, construction documents, etc.). Second, decisions must be made as to the kinds of information a schedule should show. This may be simply a list of activities to be accomplished or may include time frames, sequences, and individuals responsible.

Scheduling methods

An appropriate scheduling method must be selected and prepared in written form. While many computer programs are available to accomplish this step, care must exercised not to make the schedule overly complicated. As a communication tool, it must be presented in a clear enough manner to achieve its primary purpose. Many scheduling methods exist and each has its own strengths and weaknesses.

Full-wall scheduling

This is a somewhat antiquated method by which project tasks are listed, the individual responsible for each is noted, and a preliminary schedule prepared. The tasks are listed on 3-x-5 index cards and divided into piles for each responsible party. The list of individuals is posted on one side of an office wall, and the time frame or periods are listed along the top of the wall. All participants are assembled and the tasks are tacked to the wall based on when the activity will be started and finished. By this method, a schedule is developed and all participants understand their own responsibilities. The disadvantages of this method are many. To be most effective, all parties (including the client, consultants, contractor, etc.) must be available to participate in the planning session. On large projects, this can be an extremely time-consuming process. An advantage is that it allows a high degree of interaction between all project participants at an early stage of the job.

Bar charts

This is a time-tested and familiar device for scheduling. A bar chart shows a list of activities, the responsible party, and the duration of an activity. Bar charts are easy to prepare, are familiar to most people, and clearly communicate information (see example below in the next section). In addition, they are excellent and simple tools for monitoring. Other methods, however, are superior for planning purposes. Bar charts do have several disadvantages. For example, they do not show sequencing of events/activities and the interrelationship among tasks. In addition, on a bar chart every task appears equal in importance.

Critical path method*

The critical path method is perhaps the most widely used scheduling method. The network schedule has its ancestry in the bar chart. The inadequacies of this method resulted, in 1956, in motivating the DuPont Company to adopt the rapidly growing power of the computer for construction scheduling. The system they developed is called the critical path method (CPM).

At about the same time, the U.S. Navy developed a system called the Program Evaluation and Review Technique (PERT). The primary difference between the two is that CPM uses one time estimate and PERT uses three (most likely, optimistic, and pessimistic).

Many CPM and PERT applications in the construction process occur during the actual construction phase. The networks that are developed can range from less than 100 to many thousands of activities. It is the development of a disciplined method of planning sequences

---

* The remainder of the material in this chapter was prepared by Tom Eyerman, FAIA, and is taken from his *Financial Management Concepts and Techniques for the Architect*, 1973.

that is important in CPM and PERT, not simply the use of the computer. Not only is the planning process significant with this method, but it also encourages the effective management of schedules.

In developing a CPM schedule, the project manager must identify the interrelationships between tasks and establish their duration. He or she must prepare a schedule of activities and determine the critical paths or steps.

The first and most important step in preparing a project for networking is the correct division of the job into units of work that are relevant to the people for whom it is being prepared. For example, a network indicating just three subjects — design, working drawings, and construction — may be applicable for the design professional to show a client. For internal use, however, each of these subjects, such as working drawings, might be subdivided into architectural, mechanical, and structural drawings.

Furthermore, to be of any aid to the architectural job captain, the drawings might be subdivided into plans, sections, and elevations. Networks for internal control must be developed at the level where the work is to be performed. Once the network process is removed from the operating level, it becomes merely a theoretical tool. No matter how detailed and accurate a network may be, if it is not meaningful to the operating people, it is useless. The personal involvement of these people is what makes a network a viable tool.

The level at which the designer wants a network (commonly called the level of indenture) is determined by asking, "Who is it for?" and "What is its purpose?" Answers to these two questions are the first management decisions in developing a network.

Once the level of indenture has been established, each division of work is split into a number of component parts called activities. An activity is any definable time-consuming task necessary to execute a project. An example of an activity would be "drawing wall sections".

A network has three phases: planning, scheduling, and monitoring. Planning is the study or recognition of the events and their interrelationships necessary to complete a project. A network is a systematic attempt to plan the work. Just like a wall section shows how a wall is to be built, the network is a graphic representation of how the project will be produced in the office. The people performing this task must have a sound knowledge of how a job is put together. Moreover, they must have a thorough understanding of the particular project and of the scope involved.

Scheduling is the second phase of networking. A reasonable estimate is made of the time required to perform each event shown on the network developed in the planning stage. With the network and the estimated time for each activity, you can proceed in an entirely mechanical manner to determine the overall time required for a project. By establishing the time required, the design professional is establishing the budget for the project. Just as dimensions on a wall section tell how high the structure will be, the schedule tells how long it will take to produce the project.

Monitoring is the third phase of networking. Monitoring means nothing more than comparing what is actually being done with what was planned. In other words, "We have planned the work; now we work the plan." The network enables you to spot difficulties at an early stage of the project. This is the point where the accounting reports must tie in with the network.

The start and completion of an activity is called an event. An event is a specific point in time indicating the beginning or ending of one or more activities. An example of an event would be "wall sections completed"; whereas, an example of an activity would be "preparing wall section of exterior wall". In preparing a project for networking, it is most useful to define the last event and then work backwards from that point.

As the activities and events are developed, they are plotted on a piece of paper to create a pictorial description of the network. An activity is represented by an arrow; an event is represented by a circle. PERT is generally event orientated and CPM activity orientated.

Events and activities must be sequenced on the network under a logical set of ground rules which allow the determination of critical paths. These ground rules are found in Figure 11.1a. In example 5 of the figure, X is called a dummy variable. It represents a restraint (halting start of activity C until activity B is completed) which cannot be recognized by the conventional symbols for events and activities.

How can the networking technique represent activities which are shown in a Gantt (bar) chart as illustrated in Figure 11.1b? The answer is that there must be something which causes the decision to start an activity. The task of the person developing the network is to isolate this decision-causing point and to include it as an event on the network. In the preceding example, the decision to start construction of A is determined by approval of the design for A. Figure 11.2 shows a sample network for solution of the problem. When the network is completed, an estimate is made as to how long each particular activity will take.

A further development in networking can be to calculate the earliest and latest time an event can take place without affecting the completion of the project's final event. The difference between the earliest and latest time an event can occur is called slack time. The critical path is then simply the series of activities and events that have no slack time. The critical path, therefore, is the bottleneck route. Only by finding ways to shorten jobs along the critical path can the over-all project time

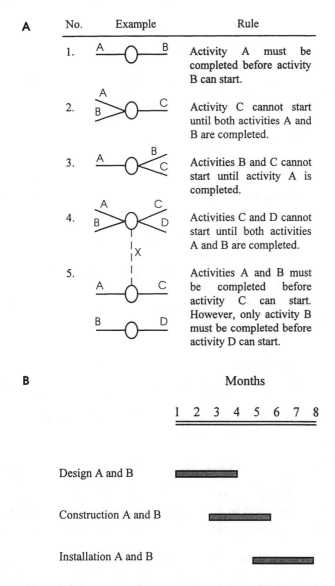

A

No.	Example	Rule
1.		Activity A must be completed before activity B can start.
2.		Activity C cannot start until both activities A and B are completed.
3.		Activities B and C cannot start until activity A is completed.
4.		Activities C and D cannot start until both activities A and B are completed.
5.		Activities A and B must be completed before activity C can start. However, only activity B must be completed before activity D can start.

B

Months

1   2   3   4   5   6   7   8

Design A and B

Construction A and B

Installation A and B

Figure 11.1   (A) Sequence of events and activities. (B) Gantt chart.

be reduced; the time required to perform non-critical jobs is irrelevant from the viewpoint of total project time. The frequent (and costly) practice of "crashing" all jobs in a project in order to reduce total project time is thus unnecessary. Of course, if some way is found to shorten one or more of the critical jobs, then not only will the whole project time be

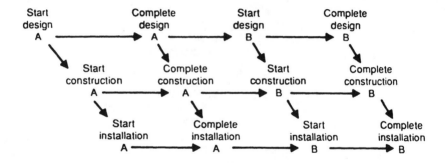

Figure 11.2   Sample network.

shortened, but the critical path itself may shift and some previously non-critical jobs may become critical.

Once the network is developed and critical path established, cost estimates are made by first determining the manpower to perform each activity. The manpower allocation is then converted to dollars to determine the direct cost of the activity. These cost estimates are used in three ways:

1. To aid the professional in determining his fee
2. To evaluate, overall, the project in terms of expenses before any action is taken
3. To provide benchmarks against which actual costs can be compared

This, then, describes the networking technique. A single example should help to clarify the process of constructing a network. The example project of building a house is taken from the article, "ABC's of the Critical Path Method," by F.K. Levy, G.L. Thompson, and J.D. West, which was published in the *Harvard Business Review* (September-October, 1963).

Shown in Figure 11.3a is a list of major tasks for building a house, together with estimated time and the immediate predecessors for each task. Following the ground rules for networking, a network can be developed as shown in Figure 11.3b. The path A-B-C-D-J-K-L-N-T-U-X is the critical path, with a maximum of 34 days.

Suppose now that October 1 is the target time for completing the project. This October 1 date is subtracted from the time for event X and the remaining time is forwarded to event S. Assuming a 6-day work period, you can see that event A must be started no later than August 23. Using the same analysis, you can see that event V may start as early as September 13 or as late as September 24.

A

Task no.	Immediate predecessors and description	Normal time (days)
A	Start	0
B	*a* Excavate and pour footers	4
C	*b* Pour concrete foundation	2
D	*c* Erect wooden frame including rough roof	4
E	*d* Lay brickwork	6
F	*c* Install basement drains and plumbing	1
G	*f* Pour basement floor	2
H	*f* Install rough plumbing	3
I	*d* Install rough wiring	2
J	*d,g* Install heating and ventilating	4
K	*i, j, h* Fasten plasterboard and plaster (including drying)	10
L	*k* Lay finish flooring	3
M	*l* Install kitchen fixtures	1
N	*l* Install finish plumbing	2
O	*l* Finish carpentry	3
P	*e* Finish roofing and flashing	2
Q	*p* Fasten gutters	1
R	*c* Lay storm drains for rainwater	1
S	*o, l* Sand and varnish flooring	2
T	*m, n* Paint	3
U	*l* Finish electrical work	1
V	*q, r* Finish grading	2
W	*v* Pour walks and complete landscaping	5
	s, u, w Finish	

B

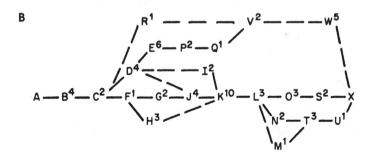

Figure 11.3 (A) Tasks. (B) Sample network based on tasks.

Benefits and limitations

The benefits of using a network are as follows:

1. A network gives the architect or engineer a method of programming the project and then a method of evaluating the office performance during the project, rather than after the project is completed.

2. A network provides a plan that can be distributed to the project team so they know where they are headed and how they are going to get there. Furthermore, a network that indicates a percent of completion for various calendar days provides the professional with a good check on the actual percent of completion. Everyone on the project team knows what must be completed by a certain date. Thus, rather than look at a completion date several months in the future, their attention is drawn to what they must accomplish in the next 2-week period.

3. A network is an excellent tool to show current and potential clients what must happen in the office to produce a set of drawings. It may help the client reach a decision if he knows the completion date may be extended a month due to his indecision; also a network can show very early in the project whether the scheduled client dates can be made. Thus, the design professional is able to give the client realistic dates of completion while avoiding the embarrassing task of explaining to an owner why the contract documents will not be finished on time.

4. The network may be used to determine what fee is actually necessary to perform the work a client demands.

5. A network gives the designer a basis to determine the future manpower requirements for the office. The networks will never give an accurate manpower projection, as small jobs will always arise that require manpower but are too small to network. With experience, however, the professional should be able to project fairly accurate manpower requirements.

6. A framework is developed which can be used to test out alternative approaches to a project.

The benefits described above do not just happen. The designer must become involved with, support, and use the network. But, in order to do this, one must:

1. Have the ability to define the project fully to a point where events and activities are clearly expressed.

2. Have the ability to keep the size and complexity of the network reasonable.

3. Have the ability to encourage employees' cooperation in developing a meaningful network.

In summary, networking cannot be delegated to a clerk; rather, it requires the active support and judgment of the project manager.

There are certain limitations of networking:

1. Networking is not a panacea for ineffective management, but it is an improvement over other project planning devices, such as Gantt charts. It also allows for the generation of analytical information that was not previously available.
2. The introduction of networking will probably increase costs. Consequently, the cost of introducing networking into a specific situation must be evaluated against the increased benefits received from the additional information that may be available. This evaluation will by no means always favor networking.
3. Networks have caused some people to look backward, seeking to place or shift the blame for a lack of progress, rather than looking forward to bring the project to a successful conclusion.

For many networking applications, simple personal computer-based software is available. This makes the use and benefits of network scheduling available to all design firms. Many excellent texts are available for guidance in developing a network schedule. One of the best is *CPM in Construction Management*, by James J. O'Brien (see the Bibliography for more information).

Other methods

Numerous other scheduling methods are in common use. The milestone chart show activities, duration, and start and finish dates. Other methods include the cumulative graph, histogram, and project chart; all have their uses and applications. Appendix D includes a selected list of scheduling computer software.

Section four

Managing the project

Chapter twelve

Project status reporting

A significant number of small design firms fail to monitor project costs adequately. This failure works to the detriment of firms in several ways:

1. Loss of control over current projects due to the inability to know the amount of money spent vs. the amount budgeted, money spent to date, and other key financial factors
2. Failure to build a historical base of information to use for future fee negotiations and project budgeting
3. Resulting inability to develop management reports to analyze problem areas in the firm
4. Lack of documentation to support claims for additional compensation

Figures 12.1 through 12.5 illustrate the development of a basic project status reporting system. The first step in developing a reporting system involves formulating a method of allocating direct labor and expenses by project. Labor is distributed by using a labor cost sheet

Figure 12.1 is prepared each time-card period and includes only labor charges made that period. Some firms alter this form by listing the phase first, then the employee. The invoice code dates when the item is invoiced (most useful in a manual system).

Project-related direct and reimbursable expenses can be recorded by the method shown in Figure 12.2. This report is normally prepared each month and includes only expenses paid (or accrued) that month. The invoice code dates when the item is invoiced.

Project summary report
Direct labor

The information contained on the labor cost sheet and the project expense sheet is used to assemble a monthly (or bi-monthly) project summary report as shown in Figures 12.3 through 12.5.

Project Name: City Hall Alterations
Project No.: 02799
Period: November 1, 1999 – November 30, 1999
Principal: Jones
Project Manager: Doe

Employee name	Phase	Hours	Rate ($)	Job cost ($)	Invoice code
Jones, R.	SD	3	12	36	
Doe, W.		2	10	20	
Clark, H.		1	8	8	
Total phase		6		64	
Jones, R.	DD	4	12	48	
Total phase		4		48	
Total project		10		112	Billed 12/2/99

Note: SD = schematic design; DD = design development

Figure 12.1 Labor cost sheet.

The direct labor portion of the report contains three sections (see Figure 12.3):

1. Current period hours and dollars from the labor cost sheet
2. Project-to-date cumulative hours and dollars from the project's inception
3. Budget as determined on the project cost plan

All labor dollars are without fringes. Principals' time and all direct labor hours are job costed at an established rate for all projects. Typically, this is accomplished by dividing annual salary (or in the case of principals, a salary draw) by 2080 hours, (52 weeks × 40 hours per week). Thus, an employee paid $20,800 per year would be job costed as follows:

$$\$20,800 = \$10 \text{ per hour} \times 2080$$

Hourly employees are generally costed at their hourly pay rate (there are many other ways of establishing job cost rates; however, this is the basic method).

Overhead

Figure 12.4 illustrates the addition of overhead to the labor spent to date. Prior to November 1, only 2 hours and $16 were spent on this

Project Name: City Hall Alterations
Project No.: 02799
Period: November 1, 1999 – November 30, 1999
Principal: Jones
Project Manager: Doe

NON-REIMBURSABLE DIRECT EXPENSES

Item	Vendor	Date	Amount ($)	Comments
Consultants	Dawson Consultants	11/05	250	
	Doe Planners	11/08	100	
Total consultants			350	
Travel	United Airlines	11/10	195	
Total travel			195	
Total non-reimbursable direct expenses			545	

REIMBURSABLE EXPENSES

Item	Vendor	Date	Amount ($)	Invoice code
Consultants	Harris Design	11/15	150	
	Stanley Survey	11/18	50	
Total consultants			200	
Printing	Robinsons Printers	11/12	25	
Total printing			25	
Total reimbursable expenses			225	billed 12/2/99
TOTAL EXPENSES			770	

Figure 12.2 Project expense sheet.

project. It was budgeted at 150% overhead (multiple of labor); hence, it is assumed that, prior to November 1, the firm's actual overhead was 150%. Thus, the prior period overhead allocation is calculated as:

$$\$16 \times 1.5 \ (150\%) = \$24.00$$

Current period overhead is calculated at $112 (current period labor) × the November overhead rate of 152%, or $112 × 1.52 = $170.24. Resulting total overhead allocation to date would be $24.00 + $170.24 = $194.24.

The current period overhead rate is not multiplied by all labor spent to date, since the overhead rate may have been different in previous periods (as in the example). Multiplying by all labor spent would be misleading for management purposes and would misstate (in this case overstate) the total overhead allocation.

Project Name: City Hall Alterations
Project No.: 02799
Period: November 1, 1999 – November 30, 1999
Compensation: $14,500 maximum[a]
Principal: Jones
Project Manager: Doe

DIRECT LABOR

Phase	Current period (hours/$)	Project-to-date (hours/$)	Budget (hours/$)
SD (including pre-design)	6/64	8/80	74/692
DD	4/48	4/48	40/280
CD			160/800
BN			10/120
CA			90/720
Total labor	10/112	12/128	374/2612

[a] From project cost plan.

Note: SD = schematic design; DD = design development; CD = construction documents; BN = bidding negotiating; CA = construction administration.

Figure 12.3 Project summary report (part one).

The overhead rate is determined by taking the firm's total overhead for the year and dividing by the total direct labor for the year (including principals' time on projects). *It is recommended that overhead be recalculated at least quarterly for budgeting and allocation purposes.* The report also shows a budget line item for overhead.

Nonreimbursable direct and reimbursable expenses

Nonreimbursable direct expenses are those cost (non-labor) items which come directly out of your fee. Reimbursable expenses are pass-through items to which you may add a markup. Figure 12.5 illustrates the complete report incorporating direct and reimbursable expenses and total spent figures. The budget figures are taken from the project cost plan, while the current period direct and reimbursable expenses are from the project expense sheet. The variation between current period and project-to-date expenses represents the previous month's expenses. The total spent figures are without profit, contingency, and reimbursable markup. At the point where project-to-date total spent (currently $1442) exceeds the budget total spent ($12,130), profit and contingency would be used to complete the project. If this situation continues long enough (if over $14,500 in total is spent), the project would be losing money.

Project Name:	City Hall Alterations	
Project No.:	02799	
Period:	November 1, 1999 – November 30, 1999	
Compensation:	$14,500 maximum[a]	
Principal:	Jones	
Project Manager:	Doe	

DIRECT LABOR

Phase	Current period (hours/$)	Project-to-date (hours/$)	Budget (hours/$)
SD (including pre-design)	6/64	8/80	74/692
DD	4/48	4/48	40/280
CD			160/800
BN			10/120
CA			90/720
Total labor	10/112	12/128	374/2612

OVERHEAD ALLOCATION (current factor: 152%)

Prior balance	$24	
Current period	$170	
Project-to-date	$194	$3918

[a] From project cost plan.

Note: SD = schematic design; DD = design development; CD = construction documents; BN = bidding negotiating; CA = construction administration.

Figure 12.4   Project summary report (part two).

Daily time sheets

A good idea from Australian architect Rob Herriot:

"I have a suggestion for your readers. Why not use daily time sheets? We changed over to daily time cards some years ago and have had a special computer program written to manage the time cost system in our office.

"Daily time cards make it even harder to lose time. This is very important from both a costing and charging point of view. The time cards are collected (or demanded) for all staff at 9:00 a.m. each morning and are immediately entered into the computer. Any extra time and effort are more than made up for by the otherwise forgotten time left off time sheets.

"With professional time billable at such a high rate, you don't have to save much time to be a long way in front. You also minimize the risk of inaccurate (or faked) time sheets and can more carefully evaluate the productivity of individual staff members."

Project Name:	City Hall Alterations
Project No.:	02799
Period:	November 1, 1999 – November 30, 1999
Compensation:	$14,500 maximum[a]
Principal:	Jones
Project Manager:	Doe

DIRECT LABOR

Phase	Current period (hours/$)	Project-to-date (hours/$)	Budget (hours/$)
SD (including pre-design)	6/64	8/80	74/692
DD	4/48	4/48	40/280
CD			160/800
BN			10/120
CA			90/720
Total labor	10/112	12/128	374/2612

OVERHEAD ALLOCATION (current factor: 152%)

Prior balance	$24	
Current period	$170	
Project-to-date	$194	$3918

NON-REIMBURSABLE DIRECT EXPENSES

	Current period ($)	Project-to-date ($)	Budget ($)
Consultants	350	400	2400
Travel	195	195	
Entertainment			
Printing		50	100
Other			100
Total non-reimbursable direct expenses	545	645	2600

REIMBURSABLE EXPENSES

	Current period ($)	Project-to-date ($)	Budget ($)
Consultants	200	400	2000
Travel		50	300
Printing	25	25	500
Telephone			200
Other			
Total reimbursable expenses	225	475	3000
Total spent		**$1442**	**$12,130**

[a] From project cost plan.

Note: SD = schematic design; DD = design development; CD = construction documents; BN = bidding negotiating; CA = construction administration.

Figure 12.5 Project summary report (part three).

This report must be accurate and must be distributed on a timely basis to project managers (within 3 business days of the end of the period). Larger firms often require more data and add options including:

1. *Previous balance for both labor and expenses.* This indicates what the project-to-date amounts were before adding the current period data. In the example shown in Figure 12.5, the previous balance for SD (schematic design) would be $80 – $64 = $16.
2. *Amount remaining to budget for each labor and expense item.* This is computed by subtracting the project-to-date figures from the budget. In the example shown in Figure 12.5, the amount remaining in the budget for SD would be $692 – $80 = $612.
3. *Percent complete by phase.* This is input by the project managers based upon their estimate of actual completion (not the amount of the budget spent).
4. *Percent of budget spent.* This is calculated by comparing the project-to-date spent by item with the budget. In the example shown in Figure 12.5, the percent of the budget spent for SD would be $80/$692 = 11.6%.
5. *Projected to completion by phase.* This assumes a linear flow of work. It is calculated by comparing the actual dollars (or hours) spent to date (by phase) with the actual percentage of completion (not the percent of the budget spent). In the example shown in Figure 12.5, assume the actual percentage of completion is 9% with $80 spent to date for SD. Hence, $80/9% = $889 at completion. Compare this number with the budget for SD of $692. If spending continues at the current rate, you will be $197 behind the budget for this phase. The difficulty with this calculation is that it assumes a linear flow of work on a project (a fact that is rarely true). In some circumstances, however, it can be a valuable indicator.
6. *Effective labor rates by phase.* This compares the average hourly rate of all employees charging time to a phase with the average budgeted rate. In the example shown in Figure 12.5 for SD, the average project-to-date rate was $10 ($80/8 hours) vs. the budgeted rate of $9.35 (692/74 hours). Therefore, the firm is using more expensive people for SD than was originally budgeted. (Considering the 9% actual completion vs. the 11.6% of the budget spent, these more expensive people are not as productive as they should be or the project is more complex than expected.) Other options should be included on the report only if they are meaningful to those managing the projects and can be provided on a regular and timely basis.

Computer networks

Design and construction organizations are slowly moving toward integration of their information management systems. The existing patchwork of databases is being merged into a centralized, computerized resource available to everyone at their own workstation. For example, project managers can receive daily, online project status reports through the implementation of daily, electronic time sheets. Drawings can be exchanged electronically between offices and with clients, consultants, and vendors. Project managers and marketers can use online databases and the Internet to obtain past project histories, general marketing data, cost data, and product information; to research codes and standards; and to exchange information, in addition to dozens of other applications.

Of particular value to project managers is the daily online availability of current/active project status. The daily collection of time charges, either on paper or through an electronic time sheet has great value to a firm and its project managers. For the first time, project managers can have a management tool, not simply a history. Daily postings improve the quality and accuracy of collected information, make the information more quickly available to those who need it, and allow greater opportunity to take corrective action on projects off budget or schedule.

Chargeable rates and profitability

There typically is no relationship between chargeable rates and profitability. Unfortunately, many design firm managers believe the opposite and strive for higher chargeable rates to no avail. Focusing on chargeable rates achieves little and frustrates staff who strive for higher rates only to see little or no impact on the bottom line. There are too many factors affecting profitability for an increase in the percent of time charged to projects to have any noticeable impact in most cases.

Firms must still aggressively charge all legitimate time to projects; however, they must still be able to bill for it and collect receivables to have any impact on profitability. As nearly all U.S. design firms are cash-based taxpayers, paper profit is irrelevant for most firms. You can charge all you want; however, if your projects have reached their maximums or if you do a poor job of managing change orders or negotiating additional compensation for scope changes, then there is no positive impact on the bottom line of your business. The exceptions are the very few firms who are able to negotiate most of their contracts on a (high) lump-sum or a straight time-and-materials basis. The reality is that most firms reach/exceed project maximums and poorly manage the change order/scope process.

Controlling project design costs

This is a primary responsibility of project managers. Many project managers continually struggle with this issue. Unfortunately, projects often end up over budget and behind schedule. Why do projects get into difficulty in the first place?

1. *Principal or senior manager meddling.* Senior managers cannot continually reassign staff based upon which project they are most interested in or because a client is pressuring them. Project managers must be assertive participants in the control of staff and other resources; otherwise, they have little chance of staying within project budgets and schedules.
2. *Poor communication with clients, consultants, and others.* Accurate information must be promptly disseminated to all who need it.
3. *Assigning staff based upon availability, not on project need or skill requirements.* Occasionally, a project is viewed as a training opportunity for young or inexperienced staff. In other cases, high workloads force the assignment of unqualified staff.
4. *Excess perfection afflicting many architects and engineers.* Architectural designers who cannot commit to a scheme, ignore budgets and deadlines, or view the program only as a rough guideline are infected with this disease. Engineers who over-detail drawings or over-specify materials also suffer from this illness.
5. *Poor scope management.* Many design firms lack an adequate system for monitoring scope of services and communicating with other team members. As a result, opportunities are lost for additional fees, and required changes are communicated to others only after additional costs for labor have been incurred. The same basic drawing may be redrawn three or four or more times.
6. *Inadequate documentation.* This ranges from poor meeting notes and minutes to design and construction change orders.

In order to help project managers meet their responsibility for controlling project design costs, there are a number of steps to be taken.

1. The project delivery system selected (see Chapter 2) must allow for effective project management free of principal/senior manager meddling. Decision-making must be pushed down to the lowest effective level in the organization through a process of cross-training and continuous improvement.
2. A great deal of homework must be undertaken before a contract is signed. The design firm and, in turn, project managers must know the overall company pricing scheme, including multipliers,

hourly rates, and any other cost data required. An evaluation needs to be performed as to the suitability of your firm to work with a particular (potential) client. For example, if you have never worked for a U.S. government agency, you have a lot to learn before signing a contract. You must also prepare a detailed scope of services. A fee budget must be established based upon the proposed scope. Schedules must not only be prepared, but followed as much as possible. Forward price your work on long-term contracts, as costs may rise over the life of the contract.

3. Communicate with all members of the project team (not only your own staff). These communications should include a regular meeting process (see Chapter 19) and procedures to inform clients and consultants when design change orders have occurred (see Chapter 14 for a suggested form). You should review your procedures, standard forms, etc. with clients and consultants before you begin work on the project so everyone knows what to expect.

4. Have an information reporting system to keep project and senior managers informed. This information must be complete, accurate, and timely. There are many commercial software packages available that capably perform this function (see Appendix D for a selected list). The availability of frequent updates is crucial. This is where a daily time sheet can be of great value.

If a project is in trouble, what can a project manager do about it? There are a number of steps to be considered.

1. Identify the problem, an idea that may appear obvious but often is not. Another project manager may need to examine a job and make some recommendations.

2. Discuss the problem with the client. Project managers should not do this for every minor issue, but certainly should do so for important issues.

3. Provide support for a project manager who is overburdened, lacks adequate staff, or is inexperienced. In rare circumstances, another new project manager may need to take over the job.

4. Change the staff if those assigned to the project are also overwhelmed or inexperienced.

5. Prepare revised budgets and schedules to reflect the new situation.

6. Examine the scope to make sure you are doing what you agreed to do.

7. Have a monitoring system in place that allows you to catch problems as early as possible.

Chapter thirteen

Team management

Personnel planning

Effective project management for small design firms requires tight control over personnel levels and use. The room for error is even less in small firms, as the loss or gain of a single project can result in great over or under-staffing and thus seriously impact firm profitability. Clearly, the hiring and firing of staff as levels of work rise and fall seriously affects productivity and requires the constant training of new people. Maintaining staff at a planned level requires not only a consistent marketing effort, but also necessitates regular forecasting of present and future workloads.

Personnel planning is not an exact science. The key to its success is the regular discipline of preparing summary plans. A computer is not essential but can help store and manipulate data. Human judgment is the most important element of an effective plan. Input to the plan should come primarily from those actually responsible for running projects; however, one individual must be responsible for putting data in final form. The plan should be prepared at least monthly and reviewed and modified as needed each week. If the firm primarily handles very small projects, it may need to prepare a plan more frequently. It is not sufficient to perform this planning exercise mentally, as this does not permit a look far enough ahead to examine workload systematically and objectively.

Figure 13.1 presents a suggested personnel planning format based upon a firm of seven technical people. Note that this firm may actually have a total staff greater than seven, but for planning purposes only technical staff or full-time equivalents are used. Full-time equivalents are part-timers converted to fractions of regular staff based on a 40-hour work week. Thus, a 20-hour-a-week intern equals one half of a full-time person. (The same calculation could be performed based upon annual hours worked using 2080 hours for an average work year without

Firm	ABC designs
Date	12/99
Current personnel level	7
Chargeable ratio	0.85
Chargeable hours per person per month	$0.85 \times 170 = 145$ (thus, 1015 available without overtime)

Project Number	Man-hours to start	1st month	2nd month	3rd month	4th month	Man-hours left
IN-HOUSE PROJECTS						
9920	1400	200	200	250	250	500
9915	450	100	150	200		
9912	1100	200	200	200	200	300
9997	200	50	100	50		
9856	2000	300	400	400	500	400
9843	1000	100	300	400	100	100
Misc.	100			50		50
Subtotal	6250	950	1350	1550	1050	1350
PROBABLE PROJECTS						
BD237	1500				150	1350
BD314					450	450
BD256					1200	1200
BD295					300	300
Subtotal	3450				600	2850
× 75%	2588				450	2138
GRAND TOTAL	8838	950	1350	1550	1500	3488
Equivalent employees		6.6	9.3	10.7	10.3	
Personnel Deficiency			2.3	3.7	3.3	
Surplus		4.0				

Figure 13.1 Personnel planning.

overtime.) If secretaries or others perform some technical work, such as specification typing (and this is charged to projects), it may be desirable to factor them into full-time equivalents.

The chargeable rate is the percentage of time actually available to work on projects (without overtime). Twelve percent of total time is normally lost to vacation, sick leave, holidays, and personal time off. In addition, when calculating a chargeable rate for total staff, a significant amount of time is normally used for overhead-related items, reducing the overall firm chargeable rate to the range of 60 to 65%. The data for calculating chargeable ratios must come from historical records. The actual hours available per employee in a month is calculated based on 4.25 40-hour work weeks per month times the chargeable rate, resulting in 145 hours available per person. For a seven-person staff, this results in 1015 hours available per month without overtime.

In general, a personnel projection should be performed for the current month plus three additional months. Beyond a total of four months, the projection becomes inaccurate except for very large projects with a predictable work flow. In Figure 13.1, the man-hours-to-start column should reflect the project manager's estimate of actual personnel time required to complete the work. It should not simply be the hours remaining in the budget, unless by chance they are the same.

The In-House Projects section reflects those projects currently under contract by the firm. The probable-projects section includes projects that are in the business-development phase, including those in the negotiation process. The hours under Probable Projects are based on a preliminary scope evaluation or on a "best guess". A weighting factor of 75% is applied to adjust for unforeseen changes. This weighting factor is for illustration only and should not be used by your firm.

Each month is totaled to arrive at the total number of hours required by all projects. This total is compared with the total available of 1015 to arrive at a total for equivalent employees. For example, in the first month, 950/145 = 6.6 employees are required, leaving a surplus of .4. As a result, the firm is properly staffed to complete its work without overtime. In the second month, the firm begins to experience a substantial shortage of staff (2.3), requiring some adjustment.

As part of the analysis of the required staff time, should be an evaluation of the skills required to perform the work. This often significantly complicates the personnel planning format shown in this chapter. While the basis principles remain the same, a detailed spreadsheet may be required to perform this analysis.

Leveling work load

Tied to forecasting personnel requirements is the need to level work load to the available staff. This can be accomplished in several ways.

1. *Overtime.* Although this is the obvious solution to short-term work increases, it can be counterproductive in the long run. Studies have shown that working 20% or more overtime for a sustained period (2 weeks or more) can result in a significant drop in productivity. This will often negate any gains achieved by working more hours.
2. *Improvement of staff productivity.* A highly productive and motivated staff can readily deal with short-term work increases. A

good working environment using tools such as CADD and other computer technology can often greatly enhance productivity or allow the examination of additional options. In addition, keeping staff motivated by providing bonuses and other incentives as well as keeping them informed can be very important.

3. *Control of discretionary time off.* A regular manpower planning process can help predict workload bulges and slack periods, allowing appropriate scheduling of discretionary time. Vacations can be scheduled for slow periods, and discretionary sick leave (for example, for elective surgery) can be anticipated and matched to workload.

4. *Hiring/firing of short-term staff.* Many firms hire some staff on a project or other short-term basis. This avoids the damaging psychology of the regular hire/fire approach by allowing short-termers to plan ahead and seek other, permanent employment while still drawing a regular salary. Some senior staff people who have left other firms prefer this approach in that they can seek their desired position while still working, even if it is at a reduced salary. Small firms often gain a great deal from having this expertise available for even a short while.

5. *Farming out work to other firms.* During very busy periods, it may be possible to shift some work to other non-competing or "friendly firms". This may be particularly true for working drawings and is often advantageous for firms with temporary increases in work or where the additional staff cannot be located or hired in time.

6. *Finding alternative activities for your staff.* When a firm's workload temporarily declines and it wishes to keep the staff together, alternatives are available. For example, many firms find marketing activities for their staff to perform. This might include the preparation of graphic materials, writing newspaper or magazine articles, performing research, etc. Alternatively, staff can be "loaned" to other firms that are busy or they may assist these busy firms by performing the work in your office.

The key to maintaining a firm's financial health is to control labor costs, which requires a regular process of personnel planning. The plan is the basis by which a firm can take prompt action to increase or decrease needed staff.

Staff management

The hallmark of a successful manager is his or her ability to delegate assignments effectively to other staff members. The growth of many design firms is inhibited by the inability of the firm's owners and managers to release the reins of control over project and firm management. As a result, crisis management prevails as the overextended individual attempts to cope with a work overload and shifting priorities.

Techniques for proper delegation

Many engineers and architects have been slow to learn the skills of delegation. Often, those individuals who try to delegate do so improperly and are likely to be disappointed by the results. Other designers, either unable or unwilling to delegate, adopt the attitude that it takes more time to explain a task than to perform it oneself. In many successful and profitable companies, managers constantly find opportunities to train staff to handle delegated activities.

Select the right task to delegate

Frequently, managers assign tasks they find undesirable to subordinates. As a result, they conduct little follow-up to see if the assignment is being performed correctly. Generally, a task should be delegated when the manager's time is more profitably spent in another activity. When selecting a task for delegation, evaluate the skill and experience of the individual receiving the assignment. If additional training is required, be prepared to devote the time and effort or do not delegate the task. Repetitive tasks are ideal for delegation. Once an assignment has been given and the task learned, the manager is free to perform more important activities.

Select the correct individual to perform the delegated task

Some tasks are delegated to individuals who find the assignment as uninteresting or unchallenging as the manager did. As a result, they either postpone the task, perhaps to a crisis point, or simply delegate it to another (third) party. This latter situation may result in the assignment being performed with inadequate supervision. The result may not be satisfactory to the original manager. In other situations, the assignment may go to the most available individual without consideration of the skills required. Often, managers delegate tasks to their already overloaded secretaries or administrative assistants. As a result, proper attention cannot be given to the assignment. A successful manager tries to match assignments with skills and time availability. He or she will try to provide the necessary training if skills or experience are lacking. When an individual is overloaded with other work, the manager should assist in developing priorities and schedules.

Give the assignment correctly

There is a tendency on the part of many busy managers simply to delegate an assignment without providing adequate instruction. The result is that the individual receiving the assignment may waste time doing the wrong task or may need to ask questions constantly. This defeats the original time-saving purpose of delegating the job. When an assignment is not explained correctly or completely, the final result may be inadequate or may lack important information. Many managers believe that taking the time to explain an assignment justifies their claim that if they take the time to explain an assignment they might as well do it themselves. This rationale fails to consider both the alternative-use value of their time and the educational benefit often gained by the individual to whom the assignment is given. You should not provide step-by-step instructions, only expectations, general guidelines, suggested resources, etc.

Set a time limit or deadline for the task

It is a wise manager who sets achievable deadlines in advance of the "real" deadline. This allows time for corrective action, if necessary, and offers the opportunity to provide additional training to the individual receiving the assignment. Failure to set a deadline implies that the assignment either is not important or can be postponed. A deadline that fails to allow opportunity for corrective action can create a crisis situation, potentially damaging to the manager's immediate effectiveness.

Provide review and control mechanisms

When delegating an assignment, it is vital to evaluate how well an individual is performing. This allows the opportunity to ask questions, conduct training, and review performance. Review sessions also provide an excellent opportunity to take or direct corrective action.

Responsibility and authority

One of the most difficult concepts for firm and project managers to implement is the equality of responsibility and authority. In many companies, project, marketing, and financial managers are assigned significant levels of responsibility. They are expected to meet budgets, deadlines, and targets, often without adequate authority to implement their decisions or to meet the needs of their tasks.

A common complaint in many small and mid-sized firms is that middle managers fail to take responsibility and assume initiative for their assignments. Often, this occurs because either responsibilities are not clearly defined or responsibilities are delegated while authority is not. Where authority is delegated, the residual right to override middle management decisions may remain with the principals, department heads, or other senior management. This problem is particularly apparent in firms where the founding principals or partners are still active and are accustomed to making all decisions. These entrepreneurs often feel the need to be highly involved in marketing, project decisions, and client meetings. As a result, they may inadvertently or by habit discourage the taking of responsibility.

Effectively delegating responsibility requires several steps:

1. *Clarify and define exactly what each individual's responsibilities include.* This requires writing a position description and developing a detailed outline of specific task assignments.
2. *Train individuals in the use of tools and techniques necessary to meet their responsibilities.* Some employees fail to assume responsibilities because they lack the necessary skills. This might be in a technical area, but more often it is due to lack of experience in communication skills, organizational techniques, or the management of people.
3. *Allow individuals the opportunity to do tasks their own way.* Unfortunately, some senior managers who have been doing a task a particular way have difficulty allowing subordinates to learn through their own errors or to develop their own way of doing things.
4. *Use restraint in supervising the work of subordinates.* Senior managers should teach, not dictate. They must also learn to encourage those with specific responsibilities to find their own solutions wherever possible. Second guessing is a sure way to destroy a subordinate's decision-making ability. As a result, he or she will fail to take responsibility for assignments.
5. *Delegate sufficient authority to perform the assignment.* In general, to perform tasks adequately requires at least roughly equal responsibility and authority. Any significant imbalance in this equation will seriously handicap decision-making ability.

Chapter fourteen

Contract management

The selection of the proper contract type can be of critical importance to the financial success of a project. Often, the contract negotiation phase is the most important stage in determining whether a project will be profitable. Design professionals generally have a particular contract type they prefer, and they should try to negotiate this type with their clients wherever possible. The following is a brief description of some of the most common contracts and the advantages and disadvantages of each.

Time and materials (hourly charges plus expenses)

This is a common type of contract, for which the scope of the work is not well defined. The design professional works at a rate that includes direct salary cost, payroll burden expense (employee taxes and insurance), general and administrative overhead (all other indirect costs), and profit to arrive at an hourly billing rate times the number of hours worked. Project expenses are billed separately, often at a markup.

This type of contract arrangement guarantees a profit to the extent that design professionals can charge for all their hours. The disadvantage of this type of contract is that few clients are willing to give the design professional such a "blank check". As a result, these contracts are usually written to include a stated maximum amount. It is important for both parties to understand whether this maximum is an estimated amount or a figure that cannot be exceeded without prior approval by the client.

Lump-sum

Contracts that provide for the design professional to perform a certain scope of work for a lump sum are widely used. This type of agreement affords protection to the client and gives design professionals a guaranteed amount for their work. Regardless of whether the work takes more or less time, the lump-sum is paid, unless there are changes to the scope of the work. In this event, a new amount is agreed upon.

Lump-sum contracts are effective for both parties so long as the design professional is experienced with the type of work required and can estimate costs correctly. It is important to include a contingency factor in the lump sum to protect the firm from the unexpected. Often, firms making extensive use of CADD prefer this type of contract as it allows them to use their system and possibly enhance profits.

Cost plus fixed fee

This type of contract is popular for government work and, in theory, should guarantee that costs are recovered and a fee or profit earned. Costs are determined by using a provisional overhead rate while the work is being performed. The actual overhead rate is then determined at the end of the contract or at the close of the fiscal year when the contract extends beyond one year. Adjustments to the provisional overhead rate are made at that time.

A problem with cost-plus-fixed-fee work in government contracting is that the government operates on an appropriations basis which ascribes certain costs to different projects. If the appropriation is used up, no further funds can be allocated to a project without considerable justification and paper work, even if the reason for the extra funds can be explained and documented. As a result, the appropriation generally becomes the spending limit on that project. Therefore, if design professionals spend less than their estimated costs, they only receive reimbursement for their actual costs. However, if they spend more, the extra costs are not reimbursed and must be covered out of their fixed fee. In theory, design professionals are still entitled to the fixed fee even if their costs are less. However, in practice, this may raise the question of whether a scope change limited their involvement, and therefore the full fixed fee may be in jeopardy. In addition to cost plus fixed fee there are variations to this contract type that provide for various incentive fee arrangements.

Multiplier times salary

This method is quite common for design contracts. The multiplier is a rate that covers payroll burden expense, general and administrative overhead, and a profit factor. It is figured by multiplying the individual's hourly salary rate times the hours worked to arrive at a billing amount for labor. Direct project expenses are billed at cost or at a markup.

The multiplier method is easy to use and assures that all costs are covered. The difference between the multiplier and the hourly rate used in time-and-materials contracts is that the latter often groups classes of employees together and bills them at an average hourly rate. A possible advantage of using average rates over a multiplier is that individual salaries are not disclosed. A disadvantage is that average rates quickly get out of date unless they are revised frequently as employees receive salary increases. This can create a situation where substantial losses occur on a project because of the variance between actual and planned hourly rates (see sidebar in Chapter 10).

Percentage of construction cost

Contracts based on a percentage of construction costs are still used by some firms. Their popularity is fading, as both clients and the design professional recognize that these contracts bear no relationship to the cost of work or the amount of creativity required in a project. Nevertheless, many firms are able to achieve satisfactory results on percentage-of-construction cost contracts and willingly accept this arrangement.

Value of service

One type of contract that is difficult to price is based on the value of the service to the client. For example, engineers who save their clients considerable money through the design of energy-efficient buildings are obviously worth more than the value of their time. Design professionals should consider the value of services in developing their pricing structure to the extent that this is possible in a competitive environment. For more information on this subject, see the discussion in Chapter 10.

There are many variations on these standard contract types, and frequently more than one type will be used. For example, until the scope of work can be clearly defined, a firm may work on a time-and-materials basis. Then, the contract may be converted to a lump-sum or other basis for the remainder of the work.

Billing and collection

Many design firms hinder the collection of accounts receivable even before signing the contract. They fail to ask the client fundamental questions concerning how they are to bill for services and how they are to be paid. In most small firms, it is the principal or project manager who should be responsible for improving the accounts receivable collection process. Unfortunately, they often do not consider regular billing and collection to be an important issue and only become concerned when the checkbook balance is low. As a result, many designers pay insufficient attention to the billing and collection process prior to or just after the contract signing. Basic questions must be answered that later will improve the collection process.

1. What is the client's payment cycle? If, for example, they regularly process invoices on the 25th and the firm does not bill until the 30th, the invoice will be delayed at least 3 weeks until the client's next payment period.
2. Does the client require a special billing form? Failure to use their form will delay payment. In many situations, the design firm may not even be informed of this reason for delayed payment for several months, thus allowing several invoices to back up.
3. Is an audit required on each invoice? Some clients, on larger projects, will require an in-house audit of each invoice. Failure to anticipate and prepare for an audit will often delay processing of the invoice.
4. Does the contract call for inclusion of supporting material, such as copies of time sheets or vendor bills, along with each invoice? Failure to enclose this backup not only will delay the invoice, but will also create additional work for the firm's support staff, who must re-assemble this material from files. Anticipating this need would allow for assembly of a billing copy during normal processing of these items.
5. Who should receive the invoice? Many clients, particularly on larger projects, separate the invoice approval and processing function from the normal project administration functions of their staff. Sending the invoice to the client's project manager, if not the approval authority, may significantly delay payment.
6. Before signing the contract, negotiate interest penalties for delayed payment. Many clients will agree to this after 30 days, and in the case of the U.S. Government, federal agencies must pay interest on past due accounts (over 30 days) if certain billing conditions are met.
7. Inform clients whom they are to contact regarding questions on invoices. It is usually best to have all questions addressed to the individual in charge of the project, who may need to consult with the bookkeeper and call the client back.

There are many other techniques to improve the collection part of the process. For example, where appropriate, a personal note from the project manager along with the invoice may help create the feeling of a personal obligation on the part of the client. This technique, however, will have little effect where the project administrative and approval individuals are different.

Regular follow-up on invoices is vital. The first contact should be made within a week to 10 days after mailing to assure its arrival and to respond to any questions. Regularly prepare and distribute accounts receivable aging reports to keep project managers informed as to the status of billings on projects.

Regular billing allows the client to perceive the firm as a business-like organization that expects to be paid on time. There is, however, no substitute for proper front-end planning and follow-through. Figure 14.1 provides a suggested billing checklist form.

In situations where the design firm is unable to collect accounts receivable promptly, there are actions that must be taken. Many firms do not negotiate interest penalties for slow-paying clients. For most well-run businesses, this would seem an obvious precaution. Unfortunately, many design firm managers believe that because interest may not be collectible, it is not worth charging. This argument is absurd, in that interest charges should be used as encouragement for prompt payment and thus becomes a device to stimulate client action. In addition, in litigation, when an interest rate has not been stipulated in a contract, the court may dictate an interest rate. With fluctuating interest rates, the court-ordered rate may not be satisfactory to the design firm.

In a recent Birnberg & Associates survey of design firms, less than one third of firms regularly charged interest on delinquent accounts receivable. The most commonly charged amount was 1.5% per month, with the range being from 1 to 2% per month. Firms most commonly began charging interest after 30 days.

Scope management

A major factor penalizing project profitability is ineffective scope management. This is the process by which a project manager has in-depth knowledge of the contractual scope of a project and effectively monitors activities that deliver the promised services.

Many practitioners complain that their clients "nickel and dime" them into insolvency. This occurs when seemingly minor client requests accumulate or expand until they significantly erode a designer's profits. In other cases, the designer is to blame. A poor job of evaluating the needs of the project during the negotiating stage can necessitate extensive work that was not anticipated.

Scope management is the process by which these activities and requests are controlled. No well-managed firm can afford not to have an effective scope management program in place. There are numerous steps required to implement this program.

Date prepared _____

Project no. _____   or Change order no. _____

Project name _____

Client name and address _____

_____

Client contact for billing questions _____

Professional contact for billing questions _____

Send invoices to (include name and address) _____

_____

Others to receive copies of invoices (include name and address):

_____

_____

General questions:

1. Client billing form required?   Yes   No (if yes, attach copy)
2. Backup required?
   All vendor and consultant invoices?   Yes   No
   Vendors only?   Yes   No
   Consultants only?   Yes   No
   Other?
3. Time-sheet copies required?   Yes   No
4. Audit required?   Yes   No
5. Normal client invoice processing date(s) _____
6. Other _____

Specific questions:

1. Fee basis?
   Multiple of direct salary expense _____
   Multiple of direct personal expense _____
   Professional fee plus expenses _____
   Percentage of construction cost _____
   Fixed amount _____
   Hourly billing rates _____
   Other (explain) _____
2. Maximum fee? $ _____
3. Reimbursable maximum (if any): $ _____
4. Errors and omissions project insurance amount to be invoiced: $ _____
5. Reimbursable markup percentages: _____ %
   All items equal?   Yes   No
   If different percentages are to be used, list them: _____

_____

6. Interest on delinquent receivables: Percentage per month: _____ %
   After how many days from the invoice date? _____ days

Client _____   Design firm _____

Approved _____   Approved _____

Date _____   Date _____

Figure 14.1 Billing checklist.

1. *Do an effective job of pre-planning on every project.* This requires detailed checklists of every possible project activity. Every design firm must have a complete and accessible database of cost and time records on past projects. This information is invaluable when analyzing a new project for potential pitfalls in the budgeting process and for comparisons with current budget and scope expectations.

2. *A system must be in place to monitor costs and time on every project.* It is inexcusable for even the smallest design firm not to have such a system in the office. Many personal computer-based systems are now available for under $1000. Manual systems are simply too unwieldy, too slow, and often too improperly designed to provide the needed information. A job-cost reporting system must allow for coding and monitoring of changes to the basic scope of services. This is often done by providing for a separate project number under the base job number. All time and costs relative to the change are then charged to this separate number. These changes are not always generated by client request and may be caused by circumstances in the design office.

3. *The individual who is in charge of the project within the design firm, must know the scope in detail.* In most offices, this person is the project manager. In small design firms, however, the principal or partners may assume this role. Unfortunately, these people may be stretched thin and may not properly monitor staff activities to ensure that they fall within the contractual scope.

4. *Changes in the scope of services must be monitored.* As mentioned before, an effective monitoring system must exist and be made available on a timely basis. It is an excellent policy to establish a separate system to record time and costs if there is any doubt as to whether the activity falls under the basic scope of services. Consolidating segregated information with the base project data is far easier than attempting to go back and separate it from the base job. Some firms even prepare budgets for these segregated change orders, although in many cases the full extent of the work required may not be known in advance.

5. *Time-sheet management is an essential activity if scope management is to be successful.* With a proliferation of change order numbers for a base project, your staff may become confused. Questions will arise as to the actual project or change that they are to charge time to. A simple process to control this requires several steps:

    a. A summary sheet listing all projects and change orders open to time charges should be given to all staff members at the beginning of each time sheet period.

    b. Every project manager must be required to review and initial each employee's time sheet before it is entered into the computer. Of course, this applies only to those staff members working on the project manager's jobs. This may be impractical in larger firms.

    c. The computer system must provide project managers with a detailed list of time charges for each of their jobs and change orders. They must examine this information carefully.

    d. Collect time sheets daily to improve accuracy.

This process must be taken very seriously. Your profit margin depends on the quality of your time-sheet management. Principals are usually the worst offenders in not completing their time sheets. To be effective, the senior people must take the lead.

6. *A communication system must be developed to inform all parties of requested or required changes to the basic scope of services.* No matter what the source of the change, all affected individuals must be notified on a timely basis. In many firms, this is done either verbally (telephone, meetings, etc.) or by lengthy letters or memos. These are often ineffective and leave a great deal open to misunderstanding. To be effective, this process must be documented, but in a quick and easy manner. Many firms develop a series of shorthand work-authorization forms that fill this need and can be sent to consultants, clients, contractors, etc. Often, these forms become part of the contract between the owner and designer and may require the client's signature before any work is actually done.

7. *Develop the ability to invoice separately for change orders.* In many firms, small changes to the project scope are absorbed as a marketing cost. However, if care is not exercised, many of these seemingly small changes grow significantly and reduce your profits. Many of these may need to be discussed on a project by project basis.

Contract change orders

The process of managing project change orders should be a relatively simple one. Unfortunately, many firms fail to prepare a system adequately to cope with these change orders (changes to base project), extras (additions to basic scope), or out-of-scope items (those whose status is not yet resolved with the client). This failure may result in lost opportunities for additional compensation, client conflicts, and problems with outside consultants.

    Several simple procedures must be established to manage the change order process. Many firms without adequate project control systems do not segregate time and expenses incurred on projects with extra or changed items. Well-managed firms have built into their job-cost accounting systems the capability of segregating changes authorized by their clients or items that may require future discussion. Typically, a suffix is added to the base project number to designate these items.

When the change is authorized by the client, and a fee agreed upon, a budget separate from the base project should be established. A regular project monitoring process must be instituted so that the project manager receives a timely status report on project changes or extras. In a situation where the manager believes an item is beyond the basic project scope, a separate job-cost accounting record must be established until the fee and scope status of this work is resolve with the client.

A general rule used by many firms is that if there is any doubt concerning the scope status of the work, a separate job-cost accounting record is automatically established. In many cases, a seemingly minor change that might normally be provided as a courtesy to the client grows into a significant cost to the firm. By immediately segregating these costs, the information is available if the firm wishes to seek additional compensation from the client. Clearly, it is much easier to segregate these costs initially than it is to attempt to pull them out of consolidated records later.

To be successful in managing project change orders, there must be an individual in charge who thoroughly understands the basic scope of services. Where no one individual is in full charge of the project, many such items may simply be lumped into the basic project cost records. Obviously, this directly impacts on the base project budget and profit potential.

Communication

The most important aspect of successfully managing project change orders is effective communication. Clients, consultants, and staff must all be kept informed of the status of any items that are either authorized or open to future discussion and resolution.

Clients

In many situations, designers and clients disagree over whether a change item or extra was authorized and how much the firm was to be paid for its work. Clients may claim that they were "only thinking out loud" and never actually authorized the designer to do the work. They may also claim that they never agreed to pay any additional compensation to the design firm. In still other situations, the designer may feel that a requested change is too minor at present (it may grow!) to request additional compensation.

To document these situations and to inform their clients, some firms have developed shorthand forms generically called work-authorization (WA) forms (see Figure 14.2). Normally, the designer completes a WA form for any activity beyond the basic scope and sends it to the client as a matter of record or for their signature. This provides notification and documentation at an early stage before extensive time and costs have been incurred. As a result, a detailed scope for the change can be established, fees determined (where necessary), method of payment outlined, schedules set, etc.

Consultants

Coordinating extras and change orders with outside consultants can be a difficult task. Since these items are not covered under the basic agreement with the consultant, questions as to scope, method of payment, amount of payment, and schedule will arise. On occasion, the consultant may proceed with work possibly beyond that required by the client or prime design firm, or the amount billed to the prime design firm by the consultant exceeds that paid by the client. One common solution is the use of a consultant work-authorization form similar in nature to that used with the client (see Figure 14.3). This form is the consultant's authorization to do the work and covers fees, schedule, method of payment, etc.

Staff

For projects with numerous change orders or extras, problems arise in keeping the staff informed of proper time and expense charges. Without accurate input, the project job cost records will be much less useful. To reduce this problem, some firms distribute copies of the work authorization forms or a summary of all active change orders to key staff members. It is their responsibility to monitor the time charges of staff working for them.

Effective management of project change orders requires a system and the discipline to follow the procedures mandated by the system. It is important to keep separate records of all extras, change orders, and out-of-scope items. Without these records, the firm will lose opportunities to obtain compensation for all actual work done and will likely have lower profits.

Work Authorization

CLIENT:	Base Job Number:_____
PROJECT DESCRIPTION:	Charge Time to:_____
LOCATION OF PROJECT:	Date issued:_____
PROJECT TITLE:	Initiated by: ☐ Client ☐ Firm Reference:

DESCRIPTION OF WORK TO BE PERFORMED:

Starting Date:_____
Est. Compl. Date:_____

Phase of Work:
☐ 0 Design; Prelim & Planning
☐ 1 Working Drawings
☐ 2 Construction & Shop Drawings
☐__ Work Order Number
☐__ Special _____

Distribution (as checked):
☐ Client contact:

☐ Design
☐ Architectural
☐ Job Captain
☐ Cost Consultant
☐ Specifications
☐ Structural
☐ Mechanical
☐ Field Superintendent
☐ Consultant
☒ Managing Principal
☒ Comptroller
☒ Accounting
☒ Main File
☐ Other_____

REMARKS

Client Approval:_____ Date:_____

_____

BILLING INSTRUCTIONS:
☐ Included in basic fee
☐ Change and/or Additional Service
☐ Not determined
☐ Special
☐ Maximum Compensation, if any $_____.

☒ Project Manager:

_____
Authorized

☒ Principal in Charge:

Project Manager to provide complete information and
check applicable boxes, including appropriate distribution
of copies.

_____
Approved

Figure 14.2 Work-authorization form.

Consultant Work Authorization

CONSULTANT:

CLIENT:

PROJECT TITLE: JOB NO._____

ESTIMATED COMPLETION DATE:_____

INSTRUCTIONS: This agreement is subject to and governed by all the terms and conditions of our agreement dated _____ entered into by the undersigned unless modified in writing.

SCOPE OF WORK:

_____

SPECIAL PROVISIONS:

_____

FEE (INCLUDING THE TERMS OF PAYMENT):

_____

Date:_____   Date:_____

Prime Design Firm   Consultant

Figure 14.3 Consultant work authorization.

Project administration

Filing project data

All projects generate huge amounts of paper with varying degrees of value. Locating a particular item can be difficult since individuals often develop their own customized filing system. Although any system will work better than no system at all, good systems have certain principles in common.

Appoint one person to be in charge of the filing system. Duplicate files and duplicate copies in project files should be avoided. Sort data into similar categories. Bind data into volumes and do not permit individual items to be separated from the group. Provide a sign-out system so that missing volumes can be located. Check frequently to be sure the file is being kept up-to-date. It is common for filing to be a free-time activity. Unfortunately, far more time is lost looking for improperly filed information than is spent in maintaining a properly planned file on a regular basis.

To make project files usable, they must be broken down into sections of manageable size. Any grouping larger than what can be placed in a single folder is too large for efficiently locating specific information. A file should not be subdivided more than is necessary to keep the subdivisions of manageable size. The volume of material accumulated on a previous, similar project is a good guide to the amount that is likely to be generated by a new project.

Avoid elaborate cross-indexing systems. Simple chronological sequencing is usually the best way to locate documents quickly, as most individuals searching a file usually have a general idea of the time a document was received or created.

Separate incoming from outgoing correspondence. Do not bother trying to file letters that answer questions raised by earlier correspondence with the earlier letter. Many letters are never answered directly, or answers to several letters may be included in a single response.

Incoming correspondence

All incoming correspondence should be routed to the person in charge of the project. Some mail may have to be re-routed to others on the project team. In most cases, it is wise to make copies of the document and distribute the copies to persons listed on a standard distribution list. Circulating a single copy to several persons is rarely successful, as the copy usually stays on the desk of the first to receive it. Distributing on a "need to know" basis can also be a problem, as the person in charge may not recognize the importance of a document to another member of the team.

To reduce the need for copying long or bulky documents, a memo or e-mail acknowledging receipt and indicating who has custody of the item can be circulated in lieu of the actual document. Zoning codes or soil-boring reports, which may be important to several team members, are often handled in this manner. It is important that the original document be added to the permanent file as soon as possible to reduce the possibility of loss.

The purpose of distributing copies is to keep the project team informed. Since the original can always be consulted, recipients should only save those documents which they will be using regularly. Technical staff members generally have no need to maintain their own job files.

Outgoing correspondence

Many firms require the project manager to sign or review all outgoing correspondence. Even seemingly routine inquiries to manufacturers may violate office policy or client requests. A copy of all outgoing correspondence should be kept in a central project file, and a separate chronological correspondence file may also be helpful. Copies should be distributed to team members in the same manner as incoming correspondence.

Interoffice/intraoffice memoranda

It is often convenient to distribute job information by memo or e-mail. Administrative information of no interest to the client is an example. Like any document related to a job, it is subject to discovery (legal) proceedings in the event of law suits. As a result, judgment must be used when preparing memos. For example, an internal memo documenting a project decision might be used to find the design firm at fault in a professional liability suit.

Job notes

Many firms find it advantageous to keep a special type of memo to record the minutes of job meetings. Most projects involve many meetings with the owner, consultants, contractors, government agencies, and others. Often, decisions reached are not clearly understood by all parties. By circulating minutes to all participants, the writer, in a sense, controls the decision-making process. Months later these minutes are invaluable for determining the understanding of the parties at the time of the meeting. If these memos are printed on colored paper and organized in a standard format, their importance will be emphasized and they will more likely be read and acted upon.

Work authorizations

During the progress of a project, many alternatives, scenarios, budgets, and schedules are discussed by members of the project team. Because most people are eager to get on with the job, they often misunderstand the significance of these discussions and proceed to do work based on preliminary assumptions. Frequently, clients do not follow through with written authorizations of work discussed. The team should expect and demand receipt of formal authorization before proceeding with work (see samples earlier in this chapter).

Telephone calls

The telephone is indispensable for communicating information and assisting in problem solving. It has the disadvantage of not providing documentation of decisions and discussions. Of course, one can tape record telephone messages but this tends to inhibit the free flow of information and may be illegal in some cases. A better technique is to record decisions reached by telephone immediately after completing the call. A telephone message pad available at each phone encourages people to record calls and decisions made. Such a message pad should be the standard 8-1/2 × 11 inches to be sure the document does not get lost in the files. This process can be invaluable if questions on a job arise at a later date.

Sales calls

Design professionals rely on salespeople to provide them with information on building materials. Salespeople regularly call on designers to obtain advance information on future sales prospects and to ensure that materials are used correctly. Designers should recognize the *quid pro quo* involved and govern their relationships with these salespeople accordingly.

Designers should maintain a healthy skepticism of anything not written in the manufacturer's published data to avoid being misled by sales enthusiasm. Most firms find it expedient to direct all salespeople to a single person (often the specifier) who is available only at specific times. This is more efficient for the firm but sometimes deprives other staff of the educational value of the sales calls.

Manufacturers' assistance

Many complicated products or systems can only be described and specified with close collaboration between the designer and the manufacturer. Designers should be careful in these situations, as this close collaboration can lead to erosion of the designer-client relationship. Clearly, the client's best interests are not always the same as those of the manufacturer, who is interested in making a sale. Particularly in publicly bid projects, the designer must carefully explain to the sales representative that the manufacturer's assistance cannot imply endorsement of his or her product at the expense of others. Even when carefully explained, bad feelings can result when a designer relies heavily on a manufacturer to prepare designs or specifications. Whenever possible, this practice should be avoided. Designers should employ a knowledgeable consultant or pay the manufacturer to design the system to avoid a potential conflict of interest.

Confidentiality

Designers work for their clients and must keep the best interests of the client in mind at all times. Most clients consider their projects to be somewhat confidential. It is important that the project manager have an understanding with the client as to what information can be released to the public or to the industry or can be used in the design firm's own marketing materials. Some owners take the somewhat unrealistic position that all information about the job is confidential and cannot be discussed with anyone outside the office.

Most clients have a somewhat elastic definition of confidentiality. Although they may not want the designer to release information to the media about the project, most clients recognize that firms must gather information from contractors and salespeople and provide them with data about the project. The best policy is to insist on staff confidentiality but to authorize the person in charge to use discretion regarding the release of information.

Dealing with consultants

Most projects employ consultants to perform certain tasks. The staff must understand the duties and limits of responsibility of the various consultants. Work authorizations should explain what is expected of the consultant. Schedules should realistically consider when the consultant can begin his or her work and when it should be completed. In addition, adequate communication is essential. Job memos and correspondence should be routed to consultants to keep them informed on the progress of the project.

Checklists

A list of items to check before completion of a project can supplement memory and provide a useful record of matters considered. Checklists take many forms and can be used with varying degrees of formality. Many organizations have developed standard pre-printed lists of activities to be performed on all projects. Unfortunately, checking such a list can become a perfunctory task and may give a false sense of security by giving the impression that items and activities not listed are unimportant. A comprehensive list by necessity includes so many inapplicable items that all items become diminished in importance in the user's mind. Some excellent checklists are published by Guidelines Publications (see the Bibliography for more information). *Masterspec* specification sections contain excellent detailed checklists tailored to the specification titles actually being used on the project.

Other standard documents can serve as useful checklists. A master specification table of contents, such as one found in *Masterformat*, published by the Construction Specifications Institute (CSI), becomes a type of checklist to remind the specifier of items that may be required by the project. Probably the most useful checklists are the informal ones prepared by the user. By jotting down a list of items to check as the project proceeds, the user has a handy reminder of specific items that need to be resolved before the job is completed. Job memoranda can also be used as a checklist. Before a project is completed, review old memos and check off all items to be sure that commitments made have in fact been kept.

Check-sets

As project teams grow, it is imperative that each member know what others are doing. Up-to-date check-sets provide this communication. But check-sets can be expensive, not only in regard to the cost of the set, but also in the time lost while tracings are being printed or plotted. There are ways to decrease the cost of check-sets and to increase their effectiveness. Provide fewer but better check-sets. Bind sets with spring fasteners and replace only the sheets that have changed since the last printing. Have several members of the team share sets. Encourage users to note incorrect or out-date information on the set. When sheets are exchanged for newer ones, the team leader can review the notes and evaluate their significance. Saving old, marked sheets can provide an audit trail when errors are encountered.

Fax machines

The facsimile machine has become as common in offices as copy machines. It is a wonderful device for speeding operations, but it also rivals the copy machine and computers in producing an avalanche of paper. Communications formerly accomplished only by the telephone must now be documented with a faxed communication.

Like telephone calls, fax messages imply an urgency that forces a quick response. This works against orderly mail processing procedures and may result in communications not being reviewed properly by the appropriate members of the project team. Many firms have a policy of following fax messages with a formal copy delivered by mail. This increases the paper blizzard and may result in a communication being answered more than once.

Many lower priced facsimile machines use a thermal type paper that does not age well. As a result, the message must be copied before filing, thereby losing some of the efficiency of the facsimile process. Fax messages should be date stamped upon receipt and routed through the same distribution channels as regular mail to ensure proper communication and response. Confidential documents should not be transmitted by fax. If a document is so important that it must be followed with a mailed copy, it should not have been faxed in the first place.

e-mail

The use of e-mail can be extremely helpful in transmitting information. It allows the recipient to read the material at his or her leisure. Unfortunately, the ease of e-mail distribution means that many individuals who do not need the information are flooded with extra reading. Firms need to develop policies regarding distribution of information via e-mail.

Project notebooks

One very successful organizational technique for project managers is the development of a project notebook for each major job. These loose-leaf binders contain project information for ready reference and portability. Their primary purposes include:

1. *Organization of information.* Project notebooks provide a ready format to organize information on projects. Typically, these notebooks contain standard sections (see Figure 14.4) as a framework to follow for every project.
2. *Reference by others.* One major benefit of project notebooks is their ready availability and easy reference by other members of the project team. This helps project managers to achieve one of their primary jobs — communicating information to others.
3. *Developing historical databases.* In many engineering and architectural offices, the most complete information is contained in the project manager's personal files. By creating project notebooks, information is organized for future reference by others,

---

Section One

List of key project contacts (client, consultants, suppliers, contractors, subcontractors, etc.) including name, firm, address, telephone and fax numbers, e-mail address, secretary/administrative assistant name, home telephone numbers; also include information such as whom to copy on correspondence, etc.

Section Two

Project scope issues, contracts, copies of change order forms, etc.

Section Three

Budget information such as design fee budgets, supporting documentation, etc.

Section Four

Construction cost data, material information, etc.

Section Five

Scheduling information and charts, such as personnel plans, Gantt, CPM, etc.

Section Six

Meeting notes, field notes, etc.

Section Seven

Additional reference material

Section Eight

Chronological correspondence file as reference

Figure 14.4 Information typically found in project notebooks. Project managers should adapt and expand this list based upon their own needs and experience. Some duplication between sections is inevitable.

including the current project manager, succeeding project managers on related projects, and marketers. This material is vital for budgeting future projects, in the event of disputes or litigation, and for general documentation and reference.

Chapter fifteen

Project cost control

Project cost control has two aspects. First is the need to control internal design costs. This requires careful monitoring of your expenditures against your fee budget. Second is the need to estimate and monitor the construction budget. Failure to do this adequately may result in exceeding a client's willingness or ability to pay for constructing the facility.

Controlling internal project costs

Perhaps the most difficult task of a project manager is staying within the project design fee budget. This is a tough challenge for even the most experienced manager. A project manager should not be alone in meeting this challenge. Many factors affect his or her ability to meet the firm's goals.

The combination of matrix management and a strong project manager has its weaknesses, but it does provide for an individual to manage and monitor the project from beginning to end. As noted in Chapter 13, to ensure its proper functioning, the matrix system must provide for an equality of responsibility and authority.

No project management system will function well without a capable staff. A high level of experience accompanied by individuals able to make quick and accurate decisions will go a long way toward keeping you within your fee budget. The goal is to achieve accurate decision-making at the lowest effective level in your organization.

Many events occur before you sign a contract that can have a significant impact on your profit potential. Specialization in one or a few types of projects allows your staff to become knowledgeable in the particular needs and problems of those projects. Research and programming materials, time, and problems can be reduced.

A poorly prepared scope of services can leave many questions unanswered. This may result in conflict with clients or require excessive unbilled change orders to meet the program. Poor scope determination can lead to inaccurately calculated fee budgets. The extra work or change orders required to overcome this problem can be very costly.

Some firms compound this error by failing to forward price their work. Contracts that last over a long period of time (a year or more) or are likely to be delayed must have an inflation clause. Without this clause, overhead increases and staff raises can eat away at the profit margin built into your project multiplier.

Communicating with clients

Many firms hurt their chances at controlling project design costs by failing to communicate adequately with clients. This failure covers a multitude of issues. Not adequately defining a scope of services leaves too many issues open for challenge or question or may result in additional unpaid requests for services. A disciplined process of recording time and expenses related to change orders is essential. Many firm managers recognize that it is far easier to consolidate information than it is to segregate after the fact.

Scheduling a client's input is essential to controlling costs. Failure to plan for this input can result in delays in decision-making. A key to making a profit on a job is to keep it moving smoothly through your office. Any delays penalize the bottom line. A regular meeting process with your client allows not only better use of your time, but also can provide a decision-making forum. Project managers also have an obligation to keep their clients informed. Communication devices such as change order documentation (see Figure 14.2), meeting minutes, regular telephone calls, etc. all help to inform clients and can contribute to better communications.

Information systems

Perhaps the most important tool needed by project managers is an information reporting system that allows monitoring of costs against the fee budget (see Chapter 12). This information should be prepared by computer. Many commercially available computer software packages exist (see Appendix D). Rarely should a firm seek to design its own software. Any claims that the commercial packages do not meet the particular record-keeping needs or method of doing business of your firm may indicate an incorrect approach on your part.

Most well-run firms today collect the time sheets at least weekly. This improves the accuracy of information and allows more current updating of project status reports. Some firms even collect and post time sheets daily. This allows an interactive process by which the project manager can use the terminal on his or her desk to check the current status of a project.

No information system is of value if the information collected is not accurate. As noted in Chapter 14, time sheets must be checked before posting. Some firms require staff to obtain a project managers' signature prior to submittal.

To control project costs, project managers must understand the information provided by status reports, and they must know how to take action based on this information. If percentage of completions is used, it must be calculated and posted as accurately as possible.

Outside consultants must also be brought into the process of controlling project costs. If they fail to meet deadlines, arrive at incorrect or incomplete solutions, or do not segregate change order information, your efforts will be affected or delayed. Wherever possible, communication processes must be established to assist in working with your consultants. If a project falls behind budget, prompt action must be taken. It is vital to catch problems as early as possible. This is especially true if your projects are of short duration where any delays in obtaining status reports can prevent effective corrective action. Staff may need to be changed in order to complete the work quickly or to correct mistakes. Time schedules and budgets may need to be revised to reflect the reality of delays or budget slippage, and the scope of services must be re-examined to ensure that you are providing what you agreed to do.

Estimating and controlling construction costs

Projects must be managed in a manner which allows for the control of all expenditures. The following examples of estimating construction costs are used with the help of data gathered and "rules of thumb". When a quick estimate is required, these methods should serve adequately, but ultimately more definitive methods must be used.

1. From past projects, cost is divided by the gross building square footage to determine the cost per square foot. In order to determine the new building budget, the cost per square foot is multiplied by the gross square footage.
2. Another method is similar to item 1 above, but, more specifically, uses past data gathered for individual building types.
3. A third method that is more specific than items 1 and 2 is to use past data pertaining to each trade to determine costs.

4. The method of determining cubic footage costs in lieu of square footage costs has advantage for projects with large gross volume areas, such as theaters and auditoriums.
5. Other rules of thumb for quick estimation of project costs are cost per unit (material), cost per bed for hospitals, etc. and per student cost for educational facilities.

Various methods which offer more sophisticated results than the "rule of thumb" methods are available for use by the cost estimator. All of these methods are dependent upon historical data, and, obviously, the more current and detailed these data are, the more reliable the estimate will be.

Some of the methods used are

1. Building unit estimating (based on unit costs of material and labor)
2. Statistical and analytical estimating (based on trends, mathematics, the use of graphs, and an overwhelming amount of information input)
3. Quantity survey estimating (based on the determination of the quantity of materials and the amount of time needed to complete specific parts of the construction)

Some methods lend themselves to earlier phases of a project, while others are required when a more detailed, concise result is needed. The estimator must have several methods at his or her disposal and must be able to determine which method is most applicable to both the type of project and the particular phase of that project in which the estimate is required.

Most of the costs of labor and material information are acquired from suppliers, contractors, and all of the other price-determining sources where costs are initiated. These data may be presented directly to the estimator or by way of publications which assemble data for the estimator who subscribes.

Many publishers of periodicals and magazines offer various types of cost-control system information. The estimator's good judgment is ultimately the determining factor as to whether or not the ongoing generation of cost analysis is maintained as accurately as possible. The human factor is not replaceable. Human error on the other hand, can be somewhat eliminated by the use of computers which not only calculate costs and analyze results, but also store cost data for use in determining construction costs.

There are additional factors that cause cost differentials in building projects, and these factors must be considered. They are the elements of costs caused by economic situations, use costs, investment costs, financing costs, operations costs, maintenance costs, and alteration and improvement costs, etc. These elements must be researched as well.

Additional estimating concepts

There are two additional aspects of project cost estimating to consider. These are life-cycle costing analysis and Value Engineering.

Life-cycle costing

Life-cycle costing analysis is a technique involving as many as a dozen areas of analysis that are examined to determine their impact on the project in terms of:

1. Capital investment costs (front-end construction costs)
2. Financing costs, including the costs of equity, short- and long-term borrowing, etc.
3. Usage costs, including people and supplies to ensure functioning of the building program (ongoing usage of the building, including staff, etc. over the life of the structure)
4. Operations cost for heating, cooling, utilities, etc. (staff, supplies, outside services, etc.)
5. Maintenance costs (staff, supplies, outside service, etc.)
6. Alteration and improvement costs (such as tenant improvements, energy retrofitting, etc.)
7. Repair and replacement costs
8. Lost-revenue costs (long-range costs of not building)
9. Denial-of-use costs resulting from delays from the establishment of need to initial occupancy

Life-cycle costing analysis is vital; in one study of a government-occupied office building over a 40-year period, the life-cycle costs were 92% for staff salaries (usage costs), 6% for maintenance and operations, and 2% for the capital costs of the building.

Value Engineering

Value Engineering is a process which identifies the general functional requirements of a building and evaluates design alternatives. This is undertaken to satisfy these requirements on the basis of the initial construction cost of the structure and its expected maintenance and operational expense over its projected life span. In essence, Value Engineering is a second look at design decisions to evaluate the balance between the most economical solution for one portion of the building against those decisions made for other parts (or systems) of the building. It involves an overview of all parts of the project and the evaluation of all implications of a design solution. For more information on Value Engineering, see Appendix B.

Chapter sixteen

Quality management

Few design firms engage in any regular or formal process of quality reviews. Some may spot check or occasionally critique designs and/or technical decisions. Rarely do designers establish a formal quality assurance program. These programs are intended to develop checking procedures, checkpoints, lines of communication, meeting processes, clarity of authority and responsibility, assignment of checking responsibilities to staff, implementation of training programs, and a wide range of other systems.

Under the pressure of meeting deadlines and maintaining budgets, most designers tend to minimize their quality assurance reviews or to ignore them altogether. Even more distressing is the lack of attention to the development of a program to improve the quality of the product being issued by the firm. The focus on quality must permeate the entire practice. Quality management is far more than simply reviewing designs or drawings at various stages of completion. It must include not only a review of the product, but also an examination of the method of operation and organization within the firm to develop designs, working drawings, and specifications.

Total Quality Management

A number of years ago, an old architect explained the difference between quality control and quality assurance. He said that quality control required three steps:

1. Making a lot of errors
2. Spending a huge amount of time and money finding them
3. Investing even more time and money to fix them (or ignore them and hope they go away)

He went on to say that quality assurance was simply a process of avoiding errors in the first place. Many of us today know this as Total Quality Management (TQM).

Unfortunately, TQM has been denigrated by many in the construction industry. Perhaps this was a result of overselling by management consultants or because of confusion arising from ISO 9000 certification (see Appendix C). Whatever the reason, TQM is a sound idea that should be incorporated in all design and construction organizations. While there are numerous philosophical and theoretical works on the subject of Total Quality Management, what is important is the practical application of theory.

Here are some suggestions as to how to apply TQM in your organization:

1. An effective project management/project delivery system is essential to controlling the quality of the product you produce. In nearly every other system, crisis management prevails, a clear point of contact is lacking for clients, no one individual is in charge of a particular project, and a variety of other weaknesses exist.

2. You should use appropriate technology, but your manual systems must be well thought out and clearly understood before automation occurs. Remember that CADD is a tool and should be part of your total system. Look toward system integration and continually seek out new applications for technology (see Chapter 12).

3. Effective communications is the basic tool for improving quality. Accurate and timely information must be conveyed to all who need it. A regular meeting process must be used. Tools such as e-mail, voice mail, faxes, etc. must be part of every project manager's daily life.

4. Project management and quality are team concepts. No one individual is capable of doing everything on other than a small project. Every team member must always think about how they can help other members do their job more effectively.

5. Cross-training is vital. This allows team members to help each other, builds support during busy times or absences, and improves the performance of everyone (see Chapter 6).

6. Every staff member's authority level must be roughly equal to their level of responsibility. This is particularly true for project managers. Without this equality, you will not properly be able to perform your job. Unfortunately, many senior managers find it irresistible to meddle in projects, particularly when they are well

acquainted with the client. This can destroy the project manager's authority and hurt project quality.

7. Every organization must strive to push decision-making to the lowest effective level. This is achieved through training, well-documented procedures, good communications, effective hiring, and a variety of other techniques. Remember, too, the three steps for making this work: your system must *permit* people to make decisions; they must be *willing* to make decisions; and they must make *correct* decisions.

8. The idea that "if it isn't broken, don't fix it" is absurd. Continuous improvement is vital. Your marketplace changes daily, the economy changes, legislation and regulations change, the labor market changes, and a thousand other things change. If you don't change, you are finished.

Peer review

One important resource that designers can call upon to improve their quality is to make use of peer review programs. These have been offered by most of the major design profession associations including NSPE, AIA, ACEC, and the Association of Engineering Firms Practicing in the Geosciences (commonly known as ASFE).

In May 1990, the *Construction Specifier* magazine reported on the history, progress, goals, and costs of peer review programs. As the magazine noted, "The ACEC first examined the concept of peer review in 1977. ACEC based much of its program on the first peer review program ever created for design professionals, established in the late 1970s by ASFE. Thirteen years later, five major organizations have endorsed the ACEC program, nearly 400 firms have been reviewed, and several types of peer review have been created."

In a 1991 Birnberg & Associates survey of nearly 900 professionals representing approximately 500 firms, less than a dozen had undertaken peer reviews. Most were not even aware of the existence of peer review programs. Many perceive peer review as a fad of the 1980s. Indeed, peer reviews are infrequent today, but this in no way lessens the value of the concept.

Why this lack of knowledge and participation? One reason may be the crisis management approach that prevails in many firms. Often bogged down in day to day project problems, many designers fail to examine alternatives to their current methods or to learn about programs that might help improve their quality.

A second reason may be the cost of peer review programs. According to the *Construction Specifier*, "The firm under review is billed for all review costs, including travel and lodging expenses incurred by the review team, administrative costs, and daily honoraria. Some smaller firms find these costs prohibitive, but compared to the costs of hiring professional consultants, the price is small."

The trade-off for the costs of the review can be great. Clearly, if the firm implements the advice of the peer reviewers, the quality of designs and drawings can be greatly improved and profitability may be enhanced. In addition, several of the leading professional liability insurance carriers offer benefits for firms undertaking peer reviews.

Components of a peer review

Firms that wish to undertake a peer review must select a review team from an extensive list of trained reviewers. All reviewers are registered

---

Peer review: help or heartburn?*

The concept of peer review is controversial in some respects. However, as designers become more vulnerable to claims, as projects become larger and more complex, as contracts become more constrictive and threatening, and as clients want more service in less time, designers should take a close look at the way their practices are functioning.

Chicago was among the first cities to establish a peer review program for architects and did so within the framework of the AIA chapter back in 1981. A quality assurance task force was appointed as a means of monitoring firms in an effort to curtail rising insurance premiums. Cook County, of which the city of Chicago is the major part, was especially hard hit with professional liability claims at that time.

The task force had the following charges:

- Increase understanding of the causes of design and building failure.
- Improve communication among architects and other members of the construction process regarding the causes and prevention of building failures.
- Elevate the quality of architectural services.
- Enhance the image of the architect as a professional able to meet the expectations of the client.
- Reduce the number of claims against architects.
- Reduce the cost of professional liability insurance.
- Reduce the number and severity of building failures and client dissatisfaction.
- Attract more architects to AIA membership.

Several meetings were held to discuss various methods of attack, to report on issues, and to plan conferences. Peer review was mentioned as a possibility, and numerous questions arose. For example, what constitutes a peer review? Who will do professionals with at least 15 years' experience, including at least 5 years in company management. There is no inherent restriction on the discipline of the reviewers because their technical knowledge is not as important as their management experience.

In a typical review, the reviewers examine documents and interview selected staff from all levels in the firm. Reviews cover six areas, including overall management, professional development, project management, personnel/human resources, finance, and business development (marketing).

Firms wishing to take advantage of peer review programs must contact their appropriate professional society. You will be provided a list of qualified reviewers. Obviously, the scheduling of a review can be complicated and can take at least 3 months to plan for an on-site visit. The review itself can be conducted and completed quickly.

---

the reviewing? Will the results be known or kept confidential? How long does the review take? What positive or negative effect might this have on the firm's practice, now or in the future?

Initial reluctance was anticipated; however, under the tutelage of Chicago attorney Paul Lurie, a leading force in promoting peer review for design professionals, it was decided that a team of professionally related individuals would conduct the initial review as a pilot program. Through interviews and examination of sample documents and resource material, they would gain an overview of the management and operational processes of the firm. From this information, the team could assess whether a firm was operating in a business like fashion and performing work in a generally satisfactory manner. The team would then make suggestions as to how the firm could improve and what methods should be considered. While general management and operational procedures were to be reviewed, financial management and technical documents were not.

In Chicago, the peer review team was composed of two or three architects and an attorney. A nominal fee was charged to cover the attorney's expenses and the expense of any clerical functions for preparation of the report. The firm could reject any reviewer it wished, for whatever reason. A letter of agreement for the review was signed by the firm and a declaration of confidentiality was signed by all reviewers. Results were kept strictly confidential and known only to the firm's managing principal. Reviewers received a preliminary copy for comment, which they subsequently destroyed. With small- and medium-sized firms, the review process took the better part of a day. A larger firm required a second day.

---

* Written by John Schlossman, FAIA.

Developing a quality assurance program*

A quality assurance program (QA) must address the quality of all services undertaken by a firm, but why should a firm take the time and make the effort to have a quality assurance program when its staff is perceived to be doing a good job? This is of particular concern, as the time spent in quality assurance is essentially nonbillable, and reluctance to undertake a QA program is understandable; however, the benefits are clear and comprise four main areas.

1. *Reduced liability.* Despite the use of great care, people do make mistakes. Anything that can be done to prevent problems and reduce potential legal claims is definitely worthwhile.
2. *Improved schedules.* Standardized methods and procedures and the more organized approaches required by a QA program usually make the work proceed faster, resulting in improved schedules.
3. *Fewer errors and omissions.* Standardized methods and procedures also result in fewer mistakes. Less-experienced staff can use tried and tested information developed by more-experienced staff people, resulting in fewer errors or omissions.
4. *Higher profits.* Performing work on a tighter schedule, producing work with fewer errors, preventing lawsuits, and cutting the time required to prepare a defense all result in higher profits.

---

*The material in this section was prepared by Jeff Orlove, AIA.

Several important questions are addressed in peer review. Is the firm organized in an orderly manner to deliver work in an efficient way? Is there proper paperwork and documentation to keep projects on track and out of trouble? What is the firm doing to identify potential problems? Does the firm provide enough challenge, satisfaction, and opportunity for staff members? Are they the right personnel to perform the work?

Prior to a formal interview, a questionnaire is sent to the managing principal outlining areas to be covered and requesting that sample materials be available for reviewers, such as a set of working drawings, specifications, front-end documents, contract forms, project logs, personnel manuals, and miscellaneous data such as conference memoranda, records of telephone calls, shop drawings, transmittals, etc.

The reviewers interview personnel at all levels within the firm, beginning with the principals and including project managers, senior architects or engineers, specification writers, construction administrators, and other staff. Questions address

A quality assurance program can also help your marketing to existing and potential clients. The program can help assure clients that their projects will be done on time and with fewer errors. A quality assurance program must include more than the construction document phase of a project. It should permeate all areas of a design practice. This includes business development, marketing, accounting, and personnel, as well as all phases of project management. A well-managed office has each of these areas clearly defined, organized, and coordinated. This allows an individual to concentrate on his or her particular area with the knowledge that other activities are being properly managed.

Steps in developing a QA program

Organization plan

The first step in developing a QA program is to create a clear and concise organization plan which should delineate the various areas of the practice. It should also clarify who in the firm has the lead responsibility for each area and how key individuals delegate other activities. In smaller offices, one person may hold more than one position, and it is extremely important to clarify roles. The plan must take into consideration strengths and weaknesses of various individuals and the needs of the firm. Both corporate and project responsibilities must be clearly identified.

Project manager system

A system for producing projects from inception to completion should be designed. Specific projects should be organized around a project

---

methods used for developing and producing documents, standardization of documents, filing and retrieval of information, reporting procedures of individuals, conducting personnel interviews and reviews, etc. The reviewers try to determine the spread of responsibility, personnel competence, and office organization.

Once the initial apprehension is overcome, the reviewers have found that everyone wants to be involved. Staffs are genuinely interested in the process and are curious to know the results. At one firm, all employees requested an informal meeting with the reviewers at the end of the day for a general discussion. They were interested in speaking out and learning how to improve procedures. The review process illustrated to them that the firm's principals were interested in improving the firm.

After the report is issued, a meeting is held with the firm's principals and the reviewers to discuss the results. Six months later, a follow-up meeting is held with the reviewers to see if suggestions and recommendations have been implemented and to learn their effect on the firm's operations.

manager. The project manager is responsible for the schedule, scope of services, fees, and organization of the team responsible for producing the design and documentation for the project.

The project manager interfaces with design team members, the production and field teams, and with the various project consultants. He or she has quality assurance responsibility for that specific project. He or she is the only project team member who has knowledge of all aspects of the job. Whether a project is a large, complicated one or a small, simple one, the same basic project management system is used. The only variable lies in the number or team members assigned to a project. The project manager may not actually conduct a quality assurance review, but he or she must plan for a review in the project budget and schedule and ensure that the appropriate reviews take place.

Quality assurance development

One major Chicago-based architectural firm has a principal-in-charge of quality assurance for both corporate and project activities. This individual works with a project QA director and a corporate QA director. The project's quality assurance director develops standardized forms, formats, methods, and procedures to be used by the various project teams. The principal-in-charge and the quality assurance project director research those methods and procedures they believe would be most beneficial. They also distribute material to the staff and conduct meetings and seminars to disseminate information that will allow projects to be done accurately and on time.

The corporate director of quality assurance deals with similar standardization of forms, formats, methods and procedures related to general operations. This part of the program includes development of a project control computer system for monitoring fee budgets for specific projects; computerization of the firm's accounting system; organization of management information systems (such as a project fee, performance, and cost database); creation and maintenance of a technical library for use by the staff; creation of a filing system for drawings, specifications, and completed project files; and development of a procedures manual for standardization of frequently performed tasks.

Lawyer as a team member

A very important aspect of a quality assurance program is the concept that prevention is better than correction. Key staff members should be encouraged to consult with attorneys on "what if" situations so that contract language can be inserted to eliminate problems. This can also help in the drafting of documentation letters to prevent a problem at a future date. Lawyers have helped develop standardized contract forms, and an attorney can also assure designers in certain situations that a perceived problem is not of concern. This allows the designer to "give" on certain business decisions without increasing liability. This often fosters greater client satisfaction.

Summary

Organization of staff is the key element in any quality assurance program. Clear responsibilities and the use of standardized forms, formats, methods, and procedures will yield the benefits of reduced liability, improved schedules, fewer errors, and higher profits. It is well worth the effort.

Chapter seventeen

Using the computer

The advent of personal computer-based graphics programs has put CADD in reach of even small offices. CADD techniques often include parts of most other design and drafting techniques. It also has some unique qualities and formidable problems of its own.

Why CADD?

Offices purchase CADD systems for various reasons:

1. *Competition.* Prospective clients often consider computer graphics capability a measure of a firm's competence and require its use. The rationale is that since everyone else is using a CADD system, we must, too.
2. *Marketing.* Computer displays can be flashy have promotional appeal.
3. *Production efficiency.* In some applications, CADD can provide startling increases in productivity.
4. *Improved accuracy.* The clarity of CADD-produced drawings is not affected by differences in drafting ability of the drafters producing the drawings. Because of the numerical precision of the computer and the semi-automatic dimensioning capability of computer graphics systems, dimensional accuracy is improved.
5. *Product uniformity.* The "style" of CADD-produced drawings is more consistent than the styles of manually prepared drawings.
6. *Additional services.* Computer systems have capabilities that enable users to provide services that cannot be done economically with manual systems.

Evaluating the need

Many users who have purchased computer graphics system have become disappointed when promised economies fail to materialize. While some applications can produce dramatic efficiencies, it requires careful management and skilled operators to make an overall improvement in productivity. Using computer graphics effectively changes the way all the work in the office is done. Real economies are related to how well the firm adjusts to the different way in which work is accomplished.

Firm managers should evaluate several points:

1. The type of work that a firm does will significantly affect the benefits of a CADD system. For example, firms with projects of a similar nature, where some repetition of details and layout is possible, will reap far greater benefits than will a firm whose client base is made up of varied and complex projects.
2. Profitability and productivity depend far more on management skill and organization than on the use of a CADD system. A well-managed firm without CADD capability will be far more productive and profitable than a poorly managed firm with CADD. "Well-managed" is not synonymous with using CADD.
3. CADD is a tool that is currently being used by nearly all design firms. Evaluate how your firm uses CADD and how it impacts your staff and organization.
4. For many firms, the interim period of establishing an initial or upgraded CADD system will be a time of lower profits and productivity. Hardware must be purchased, software obtained or written, staff trained and reorganized, and management processes developed. Unfortunately, this period may last several years until all of the proper components are in place and operating.
5. By now, most architects and engineers have discovered that productivity gains promised by vendors are overstated. If their estimates apply at all, they apply only to certain limited activities. Talk to your peers, vendors, and consultants to determine which activities offer the greatest possibility of gain.
6. Realize that you will not be able to bill for CADD as a separate item. Clients do not care how you get the work done — manually or by CADD. You will be able to bill for the person working on the CADD system at their normal hourly billing rate. Few if any clients will allow you to bill for any other CADD-related charges, except when they have made special requests for extra plotted drawings or special electronic files (tapes, disks, etc.)

There is no doubt that CADD is a valuable tool. Designers should, however, be highly skeptical of any claims or surveys that "prove" that CADD use automatically results in higher profits and productivity. There are simply too many variables to prove any such result.

Using CADD effectively

Before embarking on a CADD program, an office should carefully evaluate the character of the work it performs. The attitude of key people regarding the way they like to work and the interest of all the staff in a CADD system should be considered.

One of the keys to CADD efficiency is repetition. Any drawing task that is repetitious lends itself to machine production. It is not always easy to isolate those tasks that are repetitious. A multi-story office building does not necessarily result in a lot of repetitive drafting because normally only typical floors are drawn. On the other hand, a one-story school may contain a great deal of graphic repetition. To a machine, repetition is not necessarily repetition of large objects, but can be repetition of items as small as hatch marks. An office that specializes in a limited number of building types is likely to find more repetition than an office that does a variety of building types.

Willingness to use CADD

Computer use requires changes in the way work is done. If the firm is unwilling to change and is efficient in the use of traditional methods, CADD may not be necessary. In addition, if the staff does not understand the proper use of computers, it may not be efficient to make the investment.

Often, the staff is more ready to tackle computers than is management. Unfortunately, computer graphics cannot be implemented entirely by junior staff. The changes induced by computer graphics span all aspects of office practice. Only by continuous active support of senior management can the procedural changes required to take advantage of the computer be made. Patience is also required. It takes a comparatively short time to learn to use computer graphics equipment, but it may take a long time to learn how to combine staff and computers efficiently.

Computer use strategies

Because of the organization required and the natural separation of drawing tasks from other activities, many design firms have set up separate computer graphics departments whose primary task is to run the machines. In some cases, CADD operators may not be design professionals and may not have had previous drawing experience. The department is geared toward producing drawings based on design and analysis decisions made by others. Recent graduates may be required to spend a specific time as graphics terminal operators.

At the other extreme is an office that expects all its staff to know how to operate the equipment and use CADD whenever it is in the best interests of the project. There are, of course, many variations of these two opposing strategies. It is important to consider the implications of each strategy before devising a policy for an office.

The first strategy uses the professional draftsman approach to office organization. Since there is a great deal to learn and master in the mechanics of computer graphics operation, it is believed more efficient to have the work done by computer graphics professionals. The department, whose role is to produce drawings, can be organized with its own hierarchy. Equipment operators may have only the type of training required to run the equipment.

Such an organization can lead to serious personnel problems. Since all work filters through them, operators may act like prima donnas. In addition, the department may become divorced from the project with motivations at odds with the best interests of the project. Plus, full-time operation of graphics terminals may lead to operator fatigue and burn-out. Professionals-in-training, frustrated with the lack of project involvement, may look for ways to escape the computer graphics department.

The most serious problem with this organization, however, is its tendency to inhibit more productive use of the computer. If the operator's only purpose is to convert instructions to drawings, then he or she does not have the opportunity or motivation to take advantage of the more sophisticated use of the computer. If a computer graphics department is a separate profit center, than the motivation is to produce drawings as efficiently as possible, not to complete them in the most satisfactory manner. Professionals, not directly involved with the computer, are not aware of the computer's capabilities. Operation of the terminals being delegated to a sub-professional department tends to devalue the desirability of operating or using the computer.

The second strategy of teaching all professionals to use the system also has limitations. It takes time to learn and a certain amount of regular operation to develop and maintain one's skills. In addition, if everyone is expected to use the system as much as possible, it is likely that time conflicts will occur. Some types of graphics operations are very simple and repetitive. It is a waste of a professional's time to do them when a sub-professional can do them just as well.

On the other hand, great economies can be achieved when the decision-maker actually prepares drawings. For an engineer, having to make a sketch, send it to an operator to enter into the computer, obtain a hard copy for checking, make corrections, send it back to the operator, and then go through the process all over again is both time consuming and expensive. This is especially true if the engineer could have taken a little extra time and completed the operation at the same time the initial sketch was made. Even greater economies are possible if the engineer can combine graphics operations with analytical operations.

Most firms devise a strategy somewhere between the two extremes discussed. Their goal is to gain the advantages without the disadvantages. Firms must be aware of several other potential problems:

- *Operational problems.* Certain operational problems are endemic to computer graphics. While there are no quick fixes, awareness of the potential problems is vital.
- *Operator turnover.* It is costly and time consuming to train operators. It is disastrous if they leave the firm shortly after the training is completed, a problem that exists for all employees. Causes and solutions for the problem are probably the same for computer graphics operators as for other employees.
- *Operator burnout.* Working in front of a CRT for 8 hours is not the same as working at a drawing board. The machine has its own rhythms and demands which tend to hypnotize the operator. Regular rest periods are required and sometimes must be enforced.
- *Necessity of time away from CRT.* Operators must have time to participate in other office activities. Many offices limit time at the CRT to 6 hours a day. The other 2 hours are spent in preparation, review, and other office activities.
- *Poor morale among operators.* This is evidenced by surliness, lack of cooperation, missed deadlines, resignations, etc. There are, of course, many reasons for this, but three problems unique to computer operators stand out:
  - Operators are aware of machine capabilities which they are unable to use due to the lack of knowledge of their supervisors.
  - Unrealistic schedules resulting in excessive overtime. Computer graphics operators cannot work the amounts of overtime that young architects or engineers are often required to perform. Fatigued operators make mistakes that take more time to correct, leading to more fatigue and more errors, etc.
  - Machines are usually separated from the general work areas. Operators become divorced from other office activities and tend to develop an "us against them" mentality.

- *Machine breakdown.* Computer graphics equipment is very reliable; however, breakdowns do occur. Redundancy of equipment is desirable to prevent the office from being paralyzed by equipment failure. Offices should develop contingency plans to deal with breakdowns. Cooperative arrangements with other offices to share equipment during emergencies is a possibility. Certain kinds of machine failure can lead to loss of stored data. The loss of a disk may result in the loss of a great deal of data. While this is an infrequent occurrence, adequate backup storage is necessary. Back-up simply involves copying all data stored on a disk to an independent storage medium. Most

---

What do CADD systems do?*

In its simplest form, a CADD system is nothing more than a rather expensive drawing board. Lines and circles and arcs and text are created by the drafter and placed in a specific location by the computer. At first, computer drafting techniques seem clumsy and slow compared with manual systems. On the other hand, pushing buttons is less complicated than the pencil-handling skills acquired by the expert drafter after years of training. This has led many firms to staff the machines with untrained drafters and then teach these people to produce drawings.

While this strategy works for some applications, it assumes the drafter's primary skill is the mechanics of drawing. But machines do only what they are told. To draw a floor plan with a machine requires the same thought processes that a drafter performs when drawing manually. The machine can bisect lines, locate intersections, or compute dimensions with great speed and accuracy. But it is the operator who still determines when to bisect the line. If the drafter does not know, the machine is not going to tell him.

CADD equipment can do some tasks incredibly fast and with great precision. A drawing that took days to draw can be copied in seconds. A portion of a drawing can be mirrored or rotated or scaled differently and placed in another location almost instantaneously. These are powerful capabilities. It takes experience and forethought to know how to organize the drawings in order to use these capabilities efficiently.

With the computer, you are not restrained by the size of the paper, composition can be completely rearranged in seconds, and details can be revised beyond recognition with minor rearrangement. These qualities produce work rapidly. But they can also multiply mistakes. Good judgment, planning, and understanding of the results of actions are essential qualities of a computer graphics operator.

Computer graphics files do not contain lines and arcs. They are collections of numbers which locate the various entities in a coordinate system. Every entity has

---

* Written by Gene Montgomery, AIA.

systems provide an automatic back-up of individual files to prevent loss due to operator error. Do not confuse this with the need for back-up of the entire system. Disciplined routines by the system manager can ensure that adequate back-up measures are used. A minimum back-up program requires daily copying of the work performed (files changed) during that day. In addition, all existing files should be copied at regular intervals — at least weekly. All back-up files should be physically stored as far away from the computer as possible to guard against an accident that could destroy both the computer and the back-up files.

a coordinate associated with it and the file always "knows" exactly how far one entity is from another. All systems use this quality to produce semi-automatic dimensioning. With the computer, there is no such thing as a rough sketch. All entities are located very precisely in space. With this capability, machine drawings begin to diverge from manual drawings. A computer-generated line does not exist as a line on a piece of paper; it is a line that connects two unique points in space. The line "knows" where it is and can tell you exactly where it is any time you ask. The simplest computer graphics system has files that mimic manual drawings but also contain information not existing in manually prepared drawings.

More sophisticated computer graphic systems add information to the entity. A line can be told not only its exact position in space, but also what it is or represents. For example, a line may represent the center line of a wall. Various types of information can be encoded into the drawing and, with the proper editing procedures, reported to the user in pre-determined formats.

Not only can entities in a computer know what they represent, but they can also know they are members of a specific group. And the group may know information about itself independent of the sum of its parts. Drawings can become repositories of great amounts of information not directly related to the graphics. This file information can be edited and manipulated by drawing editing techniques which may be more easily understood than the numeric manipulation of conventional computer files.

Properties of the file can be amplified and supplemented by software and peripheral devices that use the computer files. The file can be "read" and viewed on a CRT. The drawing can be plotted by a printing device and turned into a drawing, looking much like a conventional, manually produced drawing. If the computer file was prepared as a three-dimensional drawing, software can transform it into a perspective drawing. More sophisticated software can read the drawing, analyze the information

Summary

Efficient use of people, machines, and capital is the goal in managing CADD projects. Several concepts are important:

1. The ability to accelerate a CADD project is related to the amount of equipment and number of trained operators available.
2. Many firms regularly mix manual drafting and CADD operations.
3. Plotting must be carefully planned and scheduled.
4. Intelligent personnel policies are more important than the type of equipment being used.

Changing working methods

Drawing by computer is obviously not the same as drawing manually. The time required to do a particular task is substantially different. The most efficient sequence for doing drawing tasks differs. This would not be much of a problem if a design office's major operation was drawing. But this is not the case. While staff may spend a lot of time working at a drawing board, much of that time is not spent actually drawing. An

---

associated with the entities, and reach conclusions based on that information. For example, information describing the construction of the exterior walls of the building, the windows, and their orientation can be encoded into the drawing. Software can then read this information and calculate the heat loss for the building.

Further along the path of sophistication is software that can change the graphics based on embedded characteristics. For example, the computer can be told that all 3' wide doors are now 3'4" wide, and the software would change the graphics accordingly. Despite the capabilities of even the simplest CADD system, most users seldom get past the first step of using the system as a substitute for manual drafting. The reasons for this are many:

1. All manufacturers strive to make their products "user friendly", which means that they are easy to learn and use and are tolerant of mistakes make by the user. There is a basic contradiction between tolerance of errors and the computer's innate precision. Computer language is an elaborate code with a very precise syntax. Each time error tolerance is built into the code, it becomes more complex. If the machine makes judgments about the meaning of the data entered, then potential errors are built into the data. If the machine asks for clarification, then additional time is required for data entry. The contradiction is usually resolved by requiring accurate data entry.

employee may draw for 5 minutes and then spend an hour researching a problem, talking on the telephone, or checking a shop drawing.

Offices that use the computer successfully work in a manner different from conventional organizations. With the decline in costs of computing equipment, it is feasible to provide all staff members with access to a computer. When the machines are tied together into a network (LAN, or local area network), the computer becomes much more than a machine to assist in drawing tasks.

Electronic communication among staff members, clients, consultants, and contractors can easily be performed without leaving one's desk. Drawings, sketches, renderings, and even multi-media presentations can be transmitted over networks at astonishing speeds. Networks permit the orderly and consistent backup of valuable electronic data. Research never before feasible because of time constraints becomes possible. Managers have the unprecedented capability to monitor staff performance, and all tasks are performed much faster.

All of this comes at a cost. Clerical staff may be reduced, but machines and systems require operation and support staff to keep them running. Such a staff demands higher compensation than the people they replace. As the pace of the construction process increases, there is less time for thoughtful contemplation of the issues involved.

2. Graphic data is comparatively easy to check visually. If line A is too far from line B it will appear in error and the distance can be measured and verified. But if line A is encoded to mean that it is the surface of a brick wall that is 8 feet high, then the data entry problem becomes more complex and the additional information is not visually apparent. To use the data for heat-loss calculations, you must know where the other surface of the wall is, what the internal construction is, and whether the surface is an interior or an exterior surface. A person sees and understands this construction at a glance. A computer can understand it only if the data is entered completely and accurately. It is possible to "teach" the computer rules about how to resolve ambiguous or incomplete information, but to keep the rules manageable, a great deal of standardization is required.

3. Software that performs specific tasks, such as computing heat-loss calculations, has a comparatively small number of potential users. As a result, the cost of development becomes quite high. In addition, few engineers make their calculations in exactly the same way. A small market, a non-standard process, and a variety of proprietary computer systems all conspire to make the cost of software to perform a task, which can be done quickly on a manual basis, relatively high. As a result, most users never get beyond using their machines for basic drawing work.

For example, in the past, fanciful but impractical design changes proposed by clients could be delayed by the process until it became impossible to effect the change. Now documents can be revised very quickly, the construction team can price the revision rapidly, and managers have a multitude of devices to ensure that no one delays the task. Ill-advised revisions may be constructed before more sober judgment is brought to bear.

In the computerized office, there is little use for staff ignorant in the use of the computer. Once an office starts down the road to mechanization, there is no turning back and little room for compromise. Even senior managers must learn to cope with e-mail!

Chapter eighteen

Project production techniques

The term "production" is most often used to describe the preparation of a contract or other documents. It is unfortunate that "production" is the commonly used term to describe these activities, as it implies a manufacturing process. "Communication" is a better term because it emphasizes the purpose of drawings. The purpose of all production techniques is to improve or make more efficient the communication process.

Planning

A design firm cannot design an efficient structure without advance planning. The builder of even the simplest structure needs, in his mind, a picture of the completed structure before beginning. Likewise, designers must not begin a set of plans and specifications without first deciding what instructions are needed and how they will be formulated.

In planning these instructions, consider:

1. What are you communicating, rather than what should the product look like?
2. Do you have the time and resources to do the job the same way you did it last time?
3. What problems or errors were discovered in the last job?

When preparing the project budget, make it understandable by analyzing it in as many ways as possible, such as:

1. Number of hours required per person
2. Number of hours per drawing to be produced
3. Number of hours to be spent per calendar week
4. Cost of time per task
5. Tasks for each person working on the project

Only by relating the budget to actual people and calendar dates can the budget serve as a planning tool.

Enumerate the discrete tasks to be accomplished. Units of output such as numbers of drawings often do not consider the time required for special operations. If it is necessary to spend time conferring with a code official due to local conditions, that time is not available for other production-related tasks. Failure to plan for such non-production-related tasks (which are really communication activities) can seriously upset schedules and budgets. Many offices relate time, cost, and scheduling with the use of critical path charts and diagrams. By forcing the consideration of all factors simultaneously, a more realistic plan is developed.

Traditional production

Traditionally, the only physical tools required to be an architect or engineer were a pencil and paper, T-square and triangle, and a few other drawing accessories. Drawings were the primary means of communication, and each drawing was created entirely by pen or pencil. Organization was dictated by tradition or made up by adding more drawings as the project developed. Firms can still produce drawings in this manner, but there are various techniques that improve the process.

Graphic standards

A set of drawings has a style, a vocabulary of symbols, and a method of organization that become important parts of the information transmitted. Drafters, as a part of their education and early training, learn a customary set of graphic standards. Without a formal meeting, most offices develop an informal graphic standard over a period of time. This customary dialect, if it exists, is a valuable asset for an office because it is largely self-enforcing. Most offices need to supplement this customary standard with a formal office graphic standard. New staff members can learn the office way of doing things more rapidly if they have a clearly stated standard to guide them. The act of writing down a standard requires an office to review what it is trying to communicate. Even when a strong consensus exists, an office can benefit from an occasional re-evaluation of its ways of communicating.

Various groups have also promoted industry-wide graphic standards with the idea that if all drawings used a universal language, users would better understand the drawings. It is unlikely, however, that an industry-wide standard will ever be used extensively, since many offices prefer to use their own graphic standards.

Any standard sufficiently comprehensive to guide the work of all offices adequately would probably result in some offices doing more elaborate drawings than are necessary. A standard should be tailored to the work of the office and should strive to reduce unnecessary drawing. Many customary drawings contribute little to communicating the design, and many customary symbols require more time to draw than simpler ones. In an attempt to do a good job, most drafters over-embellish their drawings and produce more than are really necessary. Standards should emphasize the minimum amount of work to be done rather than the maximum amount. Preparing a simple but comprehensive set of office graphic standards, describing what the drawings should look like and how they should be organized, is the first step in improving performance.

Mock-ups

Some project managers prepare mock-ups or mini-sets of documents (also called "cartooning") that show what drawings will be placed on each sheet of the final documents. The mock-up is drawn to scale and enables sheet composition to be determined before time is spent preparing a full-size drawing. A copy of the mini-set can be given to each member of the project team as a guide to the work to be done.

When prepared by an experienced project manager, mock-ups can save time. Inexperienced preparers tend to underestimate the number of drawings required and overestimate the space required for each drawing. Many people also have difficulty determining actual details required at the time the mock-up is created. But the attempt to locate and identify all required details produces a better selection of details to be drawn.

Standard details

All buildings are more alike than they are different. It follows that the details of construction are basically the same for most buildings. When all drawings are created with pen and pencil, much redrafting of the same detail occurs. By developing sets of standard details, an office assures that the detail will be drawn the same way each time it is used. As a result, less checking is required. Time-proven rather than one-of-a-kind details result in better construction. Standard details also guide the drafting of unique details.

It takes time to generate a comprehensive set of standard details. Most offices find it more economical to cut out and collect copies of details from previous projects. Office collections of details can be supplemented by commercially available collections such as Guidelines Publications' *Detail Tracer/Copier* book (see the Bibliography for their address and telephone). Many trade associations also publish carefully researched suggested details. By classifying and filing these standard details, an office can develop a rather comprehensive file of details in a short time. An indexing and filing system is essential. Loose-leaf binders provide the most convenient way to browse through collections of details. Similar details are placed in one binder, enabling a researcher to compare many variations of the same detail.

Some offices use the CSI *Masterformat* as a guide for organizing details. For example, window details would be filed under Section 8500. If the file became unmanageable, it would be separated into 8510 Steel Windows, 8520 Aluminum Windows, etc.

Because details are usually required to illustrate the connection of two or more primary materials, some find it more convenient to file details under broader headings, such as Wall Systems, Ceiling Systems, etc. Other systems use a keyword cross-index to locate precise details quickly.

Once prepared, a set of standard details can be used in various ways:

1. The detail can be traced onto a conventional drawing.
2. The detail can be photocopied onto sticky-back material and applied to a conventional drawing.
3. The detail can be photocopied onto clear acetate and the acetate taped to a tracing and either printed onto a sepia intermediate or photo-reproduced onto a new drawing.
4. The detail can be collected into a volume of details which supplement the full-size drawings.
5. The detail can be scanned into a computer.

With all of their advantages, why are standard details not more widely used? Some possible reasons include:

1. It takes a very carefully prepared detail to be used on another job without alterations. Most methods of reproducing standard details makes altering them difficult.
2. It takes time to locate applicable details.
3. Users tend to discount the reliability of standard details. Too often the detail does not accurately portray all parts of the construction.

Sepia overlays

When several alternate designs must be prepared, some designers draw the fixed background elements (e.g., the site plan) and then make several sepia intermediate prints of the tracing. The alternative designs are then finished on the sepia prints, thus reducing the amount of repetitive drafting required. Sometimes the process can be extended by adding subcategories of fixed elements (e.g., alternative column spacings) on different sepia background drawings and then making another generation of sepias upon which additional design work is drawn. This way, many schemes can be presented without redrawing background data. Of course, revising the sepia may take more time than was saved, if for some reason the background changes. Good-quality sepia prints are also required.

Scissors drafting

Many details have been drawn and printed in manufacturer's catalogs, reference books, and other drawings. One way of reducing drafting time is to copy these details with an office copier or just cut them out of the source document and tape them to a blank drawing sheet. A photo-reproduction firm can then photograph the pasted-up sheet and print it on Mylar or other tracing medium. Additional information can be added in the conventional manner and then the photo-reproduction can be printed along with the other tracings in the set.

A variation of this system is to prepare small drawings of repetitive plan elements and then make several photocopies of them. The photocopies can then be pasted-up like the details previously described. Some designers object to the different drawing styles likely to occur when drawings are assembled from various sources, and the cost of the photo-drawing must be deducted from the cost of the time saved.

Particularly suited for renovation work, photo-drafting is similar to scissors drafting in that photographs are printed onto drafting film. Notes and additional details can be added to the photographs the same way they are added to hand-drawn documents. By using a perspective-correcting camera, distortions on the photo can be minimized.

Pin-bar overlays

Another technique relying heavily on photo-reproduction services is pin-bar overlays. Mylar drafting film tracings are very rugged, dimensionally stable, and transparent and can be stacked several layers deep. Drawings on different layers can be seen and printed as a single composite drawing. By placing different classes of drawings on different overlays, the same tracing can be used in several different composite drawings. For example, walls can be drawn on one overlay, ceiling plan on another, title block on a third, and architectural notes and titles on a fourth. By combining the first, third, and fourth overlays, an architectural floor plan is produced. By combining the first, second, and third overlays, a ceiling plan is produced.

To keep the overlays in perfect registration, the sheets have holes punched in the borders which fit over a bar fitted with short pins; hence, the name of the system. Overlays save drafting time, but even more importantly, they can improve coordination. In multi-discipline plans, a change in one plan is correctly transmitted to the composite drawing of all disciplines.

Many tracings are produced when the pin-bar system is used. Handling and storing the tracings can be expensive and confusing. Special instructions to the printer are required, and errors in printing are frequent. As a result, reproduction costs are high and extra time for printing must be planned for. Nevertheless, the drafting efficiencies and improved coordination outweigh the disadvantages.

Details in specifications

Most details can be made to fit on a sheet of paper 8-1/2 × 11 inches. If each detail is drawn on a separate piece of paper, problems of sheet composition are avoided. The same detail may be used on different jobs. Production often goes faster when the drafter can see drawings being completed more quickly.

Some firms collect details prepared in this manner and add them to their specification book or produce a separate volume of details. Detail books can make the use of large-scale drawings more convenient by reducing the paper-handling required. Usually more details are produced if the small format is chosen. Some contractors dislike having details in a separate volume since they can get separated from the full-size drawings. This results in users trying to proceed without reference to the detail book. However, it is often more convenient to use the separate detail book than to turn back and forth between large drawing sheets.

Freehand drawings

Most designers prepare freehand sketches which are later expanded into hard-line drawings. Often these sketches are more informative and easier to understand than are the more elaborate drawings issued as working drawings. By spending some extra time on the sketch, some designers are able to skip the entire hard-line drawing process.

Freehand drawings can often more accurately portray the irregularities, misalignments, and inaccuracies of actual building conditions. Because they are drawn rapidly, they often show only the essential elements of the detail, omitting excessive information, and thus make them easier to understand.

Some designers actually sketch over hard-line drawings in order to obtain the clarity of freehand drawings. Of course, not all drafters can do freehand drawings well. But a good freehand drafter can produce a great many drawings in a short time.

Computer notes and schedules

Drawings contain many groups of notes and schedules that require hours of hand-lettering to produce. Text can be produced by a computer in a fraction of the time required to produce hand lettering. Since the text on drawings involves the distinct operations of assembling the information and then placing it on the drawing, this work can often be more economically produced by typing the data on a sticky-back appliqué which is then placed on the drawing.

Pre-printed schedule formats

Standardization can be enforced by using pre-printed schedule formats or pre-printed drawings. Pre-printed formats with blanks left to be completed by the drafter reduce the chance of forgetting vital information. Pre-printed title sheets explaining the graphic standards used in the drawings discourage drafters from inventing new symbols and conventions.

Summary

The purpose of production activities is to communicate the design concept to others. Any techniques that improve communication and reduces cost should be considered.

Chapter nineteen

Time management

Many designers continually complain about the lack of time in their lives. Crisis management in their practices makes their time management inefficient, and their families often pay the price for the designer's absence due to job demands. One important solution is reorganization of the firm's internal operations into a more efficient system. Many design firm principals attempt to manage projects day-to-day, as well as the operation of their practices. Few individuals can handle both effectively. Short of a reorganization of the firm, there are many other ideas and approaches that designers can use to improve time management.

1. *Control the number of people with whom you come in contact each day.* This can be achieved in part by holding regularly scheduled meetings. This avoids *ad hoc* meetings. Encourage decision-making within your firm at the lowest effective level. By preventing most decisions from being "kicked upstairs", bottlenecks are avoided and your time can be more effectively spent.
2. *Try using a quiet hour in your office.* The goal of this concept is to create a period of time each day for quiet office work that involves concentration. Many designers now achieve this result by coming in very early in the morning, staying late at night, or coming into the office on weekends. This strain on family life and on the individual can be lessened by using the quiet hour. Firms using this concept set aside an hour each day free of meetings, interruptions, and telephone calls. The receptionist takes messages on all except emergency calls. Intra-office calls are not permitted, and you cannot visit with anyone else in the office. Nearly all firms that have attempted to use the quiet hour have eventually abandoned its use simply because it is not enforced and individuals do not respect each other's time.

3. *Improve delegation to improve time management.* The hallmark of a successful manager is the ability to delegate assignments effectively to other staff members. Many engineers and architects have been slow to learn the skills of delegation. Often, those individuals who try to delegate do so improperly and are disappointed by the results. Other designers, being either unable or unwilling to delegate, take the attitude that it takes more time to explain a task than to perform it yourself. In many successful and profitable companies, though, managers constantly seek opportunities to train staff to handle delegated activities.

4. *Manage meetings more effectively to improve your time management.* Designers seem to love to hold meetings. This may be as a result of their desire to reach decisions by consensus or simply because the need for meetings may be inherent in the business. Whatever the reason, it is important to learn how to run an effective meeting.

Specific suggestions to improve your meetings include:

1. *Hold regularly scheduled meetings.* Rather than deal with an endless series of *ad hoc* meetings, wherever possible, hold topics to your next regularly scheduled meeting. Typically these include weekly marketing meetings, project team meetings, biweekly firm management meetings, weekly project manager meetings, and weekly or biweekly client meetings.

2. *Prepare an agenda and stick with it.* Distributing an agenda well in advance of a meeting serves as a reminder of its time and purpose, encourages advance preparation, and provides a framework for discussion. The agenda aids in preventing the meeting from going off on tangents. Agendas should have specific times listed to discuss a topic. This allows those who only need to be in attendance for a particular topic to come at the appropriate time.

3. *Apply effective time-management techniques to your meetings.* Start your meetings at the designated time even if several individuals are missing. Although a quick review may be necessary for those who are tardy, it is better to accomplish something than to sit and wait for stragglers. In your minutes, list not only those present and absent, but also those who were late and by how long (some firms fine those who are late). This latter technique is likely to be more effective for internal meetings rather than for those with clients. Discourage small talk and extraneous conversation, as they disrupt the meeting and are time wasters. Do

not allow any interruptions for telephone calls, staff questions, etc., as these also waste time and disrupt the flow of the meeting. Always set a time limit for your meetings and whenever possible stick with it by conveying a sense of urgency to the time scheduled.

4. *Use minutes and notes effectively.* The minutes not only record discussions, but they also record what decisions were made, what actions are to be taken and by whom, and deadlines that were set, and they ensure that all attendees and interested parties have a complete and accurate transcript of the meeting. In order to ensure prompt action following your meeting, distribute what is known as "instant minutes". These are nothing more than a photocopy of the minute-taker's notes. Without this immediate encouragement to actions, many individuals wait for the formal typed minutes to be distributed. This delay (often 1 to 2 weeks) may result in informal meetings to refresh memories, review decisions, etc. Formal minutes are important and should follow your "instant minutes". Some firms include an executive summary at the top of the formal minutes.

5. *Manage your meetings effectively.* Always have a chairman. Make sure everyone comes to the meeting prepared by distributing the agenda well in advance and by calling everyone to review what is expected of them at the meeting. If it is obvious that the attendees are not prepared, immediately set a new date and promptly adjourn. At the beginning of your meeting, review the agenda, summarize the results of past meetings, and emphasize the time frame allotted. At the conclusion of the meeting, review key decisions, actions to be taken and by whom, and due dates.

6. *Invite only those who need to be there to attend.* Meetings are not a substitute for effective communications. Inviting everyone who may have even a passing involvement in a project is a waste of their time. There are better ways to communicate information to others.

Other ideas

There are many other effective time-management techniques. It is important to find the ones that work best for you. For example, a pocket calendar can be helpful in planning each day and future meetings. Make a list of things to do each morning and cross them off as they are accomplished. This gives you a measure of your progress during the day.

Telephone time management

We all need the telephone. It is our most effective communications tool and source of needed information. The telephone can be a major time waster and can be disruptive to your work day. Business calls, solicitations, responses to your inquiries, etc. all interfere with the effective use of your time. Even a short call can disrupt your thought processes, creating a need to refocus on the task at hand. With some effort, you can make it less disruptive and more effective.

Suggestions

Use voice-mail systems

In a great many cases, telephone calls are made to obtain an answer to a question or to respond to someone else's question. Often, you do not actually need to speak with the other party. As most receptionists and secretaries are very poor at recording and transmitting information, the human intermediary is not always adequate.

The great advantage of voice mail is the ability to give complete and specific information to the listener. This can be a tremendous time saver and can prevent misunderstandings and confusion. Unfortunately, some firms with voice-mail systems only provide 30 to 60 seconds of recording time. Smart firms allow unrestricted time periods. The best voice-mail systems are designed so that a human being answers the telephone and connects you to voice mail if necessary. Poorly designed systems have an electronic voice answer first which offers menus or a directory of extensions (your extension should be printed on your business card).

Telephone case study

Several years ago, I attempted to telephone a friend who had recently started work at a firm in the Chicago suburbs. Calling a few minutes after 5:00 p.m., I was greeted with, "Thank you for calling ABC Architects. Our office hours are 8:30 a.m. to 5:00 p.m., Monday through Friday. Please call again." That was it. No options, no menu, no way to leave a message. I was amazed and appalled. When I finally reached my friend the next day, I told her what I thought of their unfriendly system.

Recently, I conducted a workshop which several staff members of this firm attended. I told this story and during a break privately told them that it was their firm I was referring to. They told me that it had actually cost them a major client. The client was already upset over other issues. He called one day shortly after 5:00 p.m. and was greeted with the same message. Incensed, he quickly sent off a faxed letter firing the firm. This was a very significant price to pay for a poorly conceived voice mail system.

Voice mail also allows after-hours access or communication when you are out of the office. Of course, voice mail equipment provides an excellent way to screen calls or to avoid constant interruptions while not missing an important call or information.

Many low cost voice-mail systems are available. The most common is the typical telephone answering machine. Most major telephone companies also offer voice-mail systems at a monthly charge.

Train your receptionist in proper telephone techniques

Receptionists and secretaries can cost you time when they should be saving you time. Few take a proper telephone message. A name, time, and return number are totally inadequate. They should try to assist the calling party and save you time by asking what the call concerns, offering their assistance, inquiring as to the best time for you to return the call, etc. Secretaries must learn to judge which calls are important to you and which ones can be put off. You should tell them when you are expecting an important call and what action to take when it comes in.

Do not have your secretary place your calls for you. This is highly inconsiderate of the receiving party and actually can cost you time. Most people are offended by being put on hold by a secretary until the caller gets on the line, as you are wasting their time. Keep your secretary and receptionist informed of upcoming meetings and other office activities. This allows them to suggest another individual to assist the caller in case you are not available.

Batch your return of calls

If you wish to work uninterrupted during the day, have your receptionist or secretary take messages for most calls. Set aside the last hour of the day to return these calls. When working regionally or nationwide, this can also save on long distance telephone rates (except in the Pacific time zone).

Leave complete messages

You should leave not only your name and telephone number, but also the reason for your call, the best time to reach you, and the information you are seeking (if any).

Chapter twenty

Project close-out

After the punch lists have been completed and the final actions taken, the project manager remains an important part of ensuring continued client satisfaction. Regular follow-up with a telephone call or meeting will let the client know you are concerned about service and his or her satisfaction. Some firms request client completion of a report card form (see Figure 20.1).

Time and expense spent giving attention to problems should be immediate and may not be billable but may need to be charged to marketing. The client should be kept informed of your firm's activities and should receive all relevant mailers. Remember, he or she is still your client and deserves your continued contact and attention. Some firms enter into a formal commissioning process (see Chapter 13). Others simply do follow-up to ensure client satisfaction, to learn from experience, and to obtain additional work.

Project data retention

Once a design project has been completed, a decision must be made as to the fate of the vast amount of data and material collected during the project's life. No firm can or should save everything. Many of the materials that accumulate are redundant, are working or draft versions, or are progress reports. The "pack rat" firm that saves everything will soon be overwhelmed. A firm that discards nearly everything will soon find itself regretting the decision. Equally at risk is a firm that retains the wrong materials or one that fails to organize them in a useful form. Increasingly, firms are saving much of the needed data electronically. Not only does this save space, but many clients are now requesting their own copy of this electronically stored information.

Some materials must be retained. As-built drawings should be kept for as long as the building stands. Your firm may be the only source of these important documents. Future engineers, architects, and building

Date _____   **Return to:**

                                      Bill Smith

Re: _____   ABC DESIGN

                                      123 Fourth Street

Job # _____   Albany, NY 11111

                                      Telephone (518) 666-6666

Project Manager_____   FAX (518) 777-8888

On a scale of 1 through 5 (1 being the lowest mark and 5 being the highest), how would you rate ABC on the following?

1. Listening to your needs; understanding the project   1   2   3   4   5
2. Technical capability   1   2   3   4   5
3. Innovation   1   2   3   4   5
4. Quality of our work   1   2   3   4   5
5. Our responsiveness   1   2   3   4   5
6. Frequent follow up on questions and issues   1   2   3   4   5
7. Meeting your schedule(s)   1   2   3   4   5
8. Meeting your budget(s)   1   2   3   4   5
9. Quality of our project management   1   2   3   4   5
10. Quality of our staff   1   2   3   4   5

Compared to prior years, or projects, have our services improved? Yes   No   (circle one)

Comments (if more space is needed, please use the back of this form):

_____

_____

_____

1. Were you or your client ever disappointed over something we did or did not do?   Yes   No   (circle one; if no, please explain)

2. Would you like to have a post completion meeting on this project? Yes   No   (circle one)

3. If we could do something better, or provide an additional service that would help you, what would it be? (Please list as many as needed.)

_____

_____

4. Will you consider ABC for future work?   Yes   No   (circle one)

5. May we use you as a reference?   Yes   No   (circle one)

6. Other comments or perceptions of ABC: _____

_____

If you feel ABC did an excellent job, we would welcome a "Letter of Recommendation" that we could use for our business development efforts. Thank you.

Client: _____ By: _____

Figure 20.1 ABC Design report card form. (Adapted from Baltes/Valentino Associates, Ltd.; Phoenix, AZ; 602-374-1333.)

owners will be saved much time and expense if they have these drawings available. The project specifications are also vital to enable others to review the material and equipment used.

For historically significant buildings, additional information such as preliminary design drafts may be retained. This will allow future historians and preservationists to consult documents that indicate the designer's thought processes. However, the determination of historical significance is for the future to decide, not the ego of the designer. One clue though, is the original purpose of the building. Single-purpose buildings, especially for a branch of government or for a single corporate user, often have the greatest initial impact and the longest life span.

Legal concerns

Usually, the single overriding factor in determining the retention of data is concern over the possibility of future lawsuits. Most states have statutes of limitations for the filing of legal actions in construction projects. In general, these statutes will dictate how long you should retain certain information. In most cases, 10 years will be adequate. Check the statutes in the states where you work.

Typically, materials retained because of legal concerns include a final set of drawings at the completion of each project phase. The owner-designer contract and amendments (change orders, etc.) must be retained permanently. Correspondence that could be used to pinpoint responsibility or a standard of care must be saved. Many attorneys would advise keeping all correspondence until the statute of limitations expires. In addition, job notes, diaries, and field inspection reports must be retained.

Management and marketing use

Aside from legal and historic uses, the most important reason for reviewing and organizing project material lies in its subsequent management and marketing applications. Many engineers and architects are contacted by owners years after the original project completion. This can be the source of much additional billable work and requires ready access to past project data. Obviously, it is essential to save and organize critical information to allow for timely response to client needs.

Project managers and marketers

In many design firms, the project managers have the most complete and accurate files. Often, essential data never make their way to the main files of the firm. Project managers may leave the company, taking irreplaceable records with them. Even when they remain with the same firm, the information they have is rarely available to other project managers or to the marketing staff. If the main files have gaps, then much time is wasted seeking material, records, reports, etc. that should be readily available.

Your marketing staff has a continual need for historical data on projects. Not only must the appropriate information be retained, but it must be organized to allow for quick retrieval. Without this organization, a great deal of potentially productive time can be wasted searching for information on completed projects.

For example, many proposals require submission of information on past similar projects. This often includes fees, consultants used, owners' names and addresses, contractors and subcontractors involved, staff and project managers' names, etc. Some of this may be easily recalled on significant recent projects, but often this is not the case.

In an effort to organize and ensure comprehensive records, some firms have established completed project files. These provide a checklist of required information and centralize all essential historical material (see Figure 10.1 for a sample). They are not a replacement for the firm's main files. The completed project file serves as a supplement, containing key data of value to other project managers and marketers.

Bibliography

Bibliography

Books and manuals

1. Birnberg, Howard, *Financial Performance Survey for Design Firms*, Birnberg & Associates, Chicago, annual survey.
2. Birnberg, Howard, *New Directions in Architectural and Engineering Practice*, McGraw-Hill, New York, 1992.
3. Clough, Richard H. and Glenn A. Sears, *Construction Project Management*, 3rd ed., John Wiley & Sons, New York, 1991.
4. Cottman, Jr., Ronald, *Total Engineering Quality Management*, American Society for Quality Control, Milwaukee, WI, 1993.
5. Hart, Roger D., *Quality Handbook for the Architectural, Engineering and Construction Community*, American Society for Quality Control, Milwaukee, WI, 1994.
6. Haviland, David, *Managing Architectural Projects: The Project Management Manual*, The American Institute of Architects, Washington, D.C., 1984 (out of print).
7. Ingardia, Michael and John F. Hill, *Contracting for CADD Work: A Guide for Design Professionals*, American Consulting Engineers Council, Washington, D.C., 1997.
8. Kerzner, Harold, *Project Management: A Systems Approach to Planning, Scheduling and Controlling*, 6th ed., VNR, New York, 1994.
9. Levy, Sidney M., *Project Management in Construction*, 2nd ed., McGraw-Hill, New York, 1994.
10. O'Brien, James J., *CPM in Construction Management*, 4th ed., McGraw-Hill, New York, 1993.
11. Pinto, Jeffrey K. and O.P. Kharbanda, *Successful Project Managers: Leading Your Team to Success*, VNR, New York, 1995.
12. Ritz, George J., *Total Construction Project Management*, McGraw-Hill, New York, 1994.
13. Stephenson, Jr., Ralph, *Project Partnering for the Design and Construction Industry*, John Wiley & Sons, New York, 1996.

14. Stitt, Fred, *Production Systems for Architects and Engineers,* McGraw-Hill, New York, 1993.
15. Stitt, Fred, *Architect's Detail Library,* VNR, New York, 1991.
16. Warne, Thomas, *Partnering for Success,* American Society of Civil Engineers Press, New York, 1994.
17. Woodward, Cynthia, *Human Resources Management for Design Professionals,* The American Institute of Architects, Washington, D.C., 1990.
18. *AIA Handbook,* The American Institute of Architects, Washington, D.C., current edition.
19. *Project Management Survey for Design Firms,* Association for Project Managers, Chicago, IL, annual survey.
20. *Style and Libel Manual,* 6th ed., Associated Press, New York, 1996.

Other resources

American Society for Personnel Administration (ASPA), 606 North Washington Street, Alexandria, VA 22314; (703) 548-3440.

American Society for Training and Development (ASTD), 1630 Duke Street, P.O. Box 1443, Alexandria, VA 22313; (703) 683-8100.

Association for Project Managers (APM), 1227 W. Wrightwood Avenue, Chicago, IL 60614; (773) 472-1777.

Construction Specifications Institute (CSI), 601 Madison Street, Alexandria, VA 22314; (703) 684-0300.

The Guidelines Letter, Guidelines Publications, P.O. Box 456, Orinda, CA 94563; (415) 254-0639.

University of Wisconsin-Engineering Professional Development, 432 North Lake Street, Madison, WI 53706; (608) 262-2061.

Zweig-White & Associates, One Apple Hill, Box 8325, Natick, MA 01760; (508) 651-1559 (publishers of numerous construction industry surveys).

Appendices

Appendix A

Project delivery methods

In the construction process, there are three primary methods of project delivery which include:

1. Traditional straight-line
2. Fast-track
3. Design-build

No one method is superior in itself to the others. The primary factors that influence the use of one method over another include time frames, cost constraints, and quality. Other project delivery methods could be identified; however, many are only variations of those listed above.

Traditional straight-line

The process begins with the designer determining the client's (owner's) physical and functional needs. The designer then follows through with creation of the building design, obtaining the necessary project approvals, producing all drawings, obtaining competitive bids, or negotiating prices with contractors before construction begins (see Figures A.1 and A.2). This flow allows complete development of the design program, fully developed design solutions, and complete pricing of building costs prior to construction.

It is this method that is most familiar to designers in that it blends most smoothly into their operations. This method allows fairly simple planning of personnel needs and the use of traditional management methods. Generally, by using this method of project delivery, the designer has greater input into the project program than with the other methods.

Figure A.1 Project delivery systems: basic elements.

For private-sector owners, pricing is often based upon a lump-sum general contract, under which the contractor constructs the client's building for a fixed price based upon the designer's drawings and specifications. The contractor will have (or should have) performed a complete analysis of costs for the project, including subcontracts, special equipment requirements, materials, etc. If the contractor has properly completed his estimates, a residual should remain upon completion of the building. In the event that he exceeds this fixed amount, there usually is no recourse other than renegotiation of the contracts.

It is not uncommon for the client and/or designer to retain outside cost consultants to monitor project costs during design or to monitor the accuracy and fairness of the bids or the contractor's negotiated price. The deliberate nature of the traditional straight-line method allows opportunity for the client to increase the budget if necessary, to initiate redesign of the project, or to reduce the project scope. Particularly with public sector owners, this process is vital in that often the budget is set by appropriation, and it is not always easy to obtain additional funding.

If the contractor is locked into a lump-sum contract and the project costs exceed this amount, the contractor is subject to a loss. Many prefer the use of cost-plus contracts. In this method, the contractor is reimbursed for all project costs and is paid a fee in addition. A pure cost-plus contract reduces the contractor's incentive to keep project costs down; hence, the use of upset maximums (at a percent over the original estimates) is widespread. Beyond that point, the financial burden of cost overruns shifts back to the contractor. Often, however, there is an

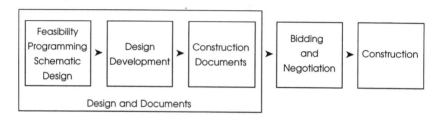

Figure A.2 Project delivery systems: traditional straight-line.

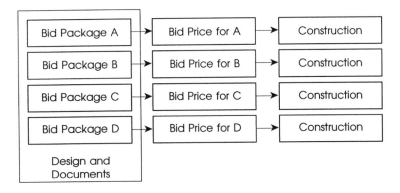

Figure A.3 Project delivery systems: fast-track.

agreement between the owner and contractor to split all savings under the upset as an incentive for the contractor to reduce costs.

Fast-track

With quickly changing economic conditions and needs, it is not always possible to wait for completion of a building under the traditional straight-line method of project delivery. In response, the fast-track method was developed to allow for the overlap of the design and construction phases. Upon completion of the project program and of the schematic design, the project is divided into bid packages that are awarded (or negotiated) in order by construction logic so that the earlier packages can be in construction while later packages are still in design (see Figure A.3).

The effect of the fast-track method on designers can be great. It increases the demand for individuals with specific skills in that the needs of construction imperatives and costs require early decision-making in key areas (such as mechanical systems, which often require long lead-times for their manufacture). The effect of these decisions on later bid packages must be anticipated. The input of the building product manufacturer or supplier to the design firm, the general contractor, and the construction manager is often key to the success of this method. This input may be in the form of evaluating product performance, costs (initial and life cycle), delivery times, etc. Often, contracts are let on an installed basis with a performance specification, again requiring a manufacturer's input.

The use of multiple bid packages (contracts) and the general complexity of many projects have required a construction manager (CM) to ensure the performance of the contractor(s) and the accuracy of both cost estimates and schedules. In this sense, the CM performs the function

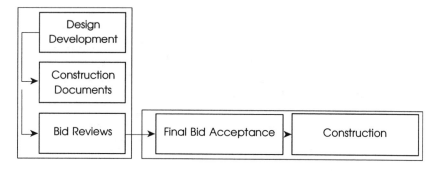

Figure A.4   Project delivery systems: design-build.

carried out by the general contractor in less complex projects. As mentioned earlier in the text, the CM often functions as the general contractor; however, any qualified party, whether the architect, engineer, etc., could be the CM.

Design-build

The use of the term "design-build" has come to be an umbrella covering many variations of the same process. Basically, this method of project delivery involves quoting the owner a price at an early stage for both design and construction of the project (see Figure A.4).

Given the limited nature of the building program at this point, there is a great burden placed on the design-build firm to be able to identify general building components and costs accurately. The essential identifying element of the design-build method of project delivery is the single point responsibility of the design-build principal. (This principal may be the architect, engineer, general contractor, etc.).

Often, there is a great number of variables in the selection of materials, equipment, and systems. Such variables offer many excellent opportunities for the design-build team, as any cost savings achieved during construction are beneficial to the design-build contractor.

Appendix B

Value Engineering*

Introduction

The initial impetus for Value Engineering (VE) in the design and construction industry came from Federal government agencies in the 1960s. Thirty years ago, Value Engineering was viewed as a way of reducing over-budget project costs. The old Value Engineering focused on cost reduction, frequently at the expense of client desired project features.

Today, a Value Engineering study seeks to balance client-required project performance with the costs necessary to achieve it. This changed focus allows project managers to offer a separately identifiable service and to involve client and building team representatives in an intimate, exciting, and time-limited process. This builds team camaraderie, deepens project understanding, and opens communications between project team members.

Modern Value Engineering fits right into the proactive practices of partnering and Total Quality Management (TQM) by strengthening project team relationships, empowering decision-making at the project team level, and identifying criteria by which customers (project owner and users) will judge the success of the finished project.

A successful Value Engineering study has four principle characteristics:

- Implemented results
- Customer involvement
- Thorough Function Analysis
- Strict adherence to the VE job plan

---

* Written by Howard Ellegant, AIA, CVS, who is an architect and Certified Value Specialist with a practice in Evanston, IL, and is devoted to Value Engineering consulting in the design and construction industry (telephone 847-491-0115).

Implemented results

Implemented VE team recommendations are a primary measure of VE success. To ensure full benefit of team recommendations, the owner and designer must buy into them. The surest way of achieving this is to actively involve them in the VE process.

To "ROAR" down the implementation path requires Responsibility for implementation, Ownership of the recommendations, and Authority and Resources to implement the VE study team's recommendations. An empowered project team can have all of these ingredients, and it is the project manager that coordinates them. The biggest stumbling block to ROARing ahead with implementation is lack of ownership of the recommendations.

Still practiced today, the old VE uses a peer review team to perform an objective VE study of the design team's work. An immediate consequence is to establish an adversarial relationship between the design team and the VE study team. The owner is caught in the middle and must decide between the VE team's recommendations arrived at over a short period of time and the design team's often defensive reaction to them. The very people who have to approve and implement the recommendations have no ownership of them and no stake in a positive outcome!

People do not like change, but if they participate in creating it they understand and own it — and can ROAR their approval and acceptance! Using the project team as the VE study team dramatically increases implementation by adding the vital factor of ownership to responsibility, authority, and resources. The VE process, specifically thorough Function Analysis facilitated by a knowledgeable VE specialist, provides the necessary objectivity for project review.

Involve the customers

The old Value Engineering too often focuses on the engineering and does not pay enough attention to the value. I define value as the relationship between customer project acceptance and project cost (value = acceptance/cost). Acceptance, in turn, is composed of customer expectations and how well the project meets them. Cost includes initial capital plus follow on costs — total life-cycle costs. While designers influence costs through their responses to customers' requirements, the customer is the final judge of good or poor value of the result.

The old VE concentrates on driving project costs down through cost-saving recommendations to change scope, materials, and systems — often with little or no regard for owner acceptance. Modern VE also suggests cost-saving changes but does so within a context of customer project acceptance criteria and their relative importance to each other. Seeking a balance of project performance and costs results in cost savings as well as re-investing some of the savings to increase project acceptance.

At the start of a VE study, we conduct a focus group to solicit from the customers (owner, users, operators, etc.) their criteria for a successful project. The VE team uses this information to determine where to concentrate its effort and generate alternative solutions to improve project performance and reduce costs by correcting problems the customers highlight.

Modern VE uses a multi-discipline team of project stakeholders to break down the design into functional performance elements. Costs and benefits are assigned to each element. The value of each functional performance element is measured by comparing its costs and benefits. Total project value is improved by making appropriate changes to balance performance and cost. Owners and designers play an integral role in defining project performance and generating and evaluating alternative ideas based upon the technical feasibility, cost impact, and political acceptability of each idea. Modern VE gives the project manager the opportunity and the tools to manage both project cost and owner satisfaction.

Function Analysis

The process of breaking the design down into performance elements is called Function Analysis. It is a systematic approach to identifying and analyzing what customers *need* and *want* from facilities to support their operations. Function Analysis is the foundation of VE and is crucial to a successful study that proposes changes to improve project value, not just reduce cost — changes that do not detract from performance or aesthetics but enhance both.

As a purchasing agent during World War II, Lawrence D. Miles, the originator of VE, specified functions to suppliers and let them suggest what materials, parts, or assemblies could be used to do a particular job instead of the ones specified by the engineers, many of which could not be had at any price because of war-time scarcity. Miles was using performance specifications, but these were performance specifications with a difference. They focused on the reason a part existed rather than its longevity, strength, elasticity, or other engineering characteristics. Miles' performance characteristics (functions) were identified by understanding how the part would be used — the job it had to do for the customer.

Miles also reasoned that the cost of accomplishing functions could be derived by careful analysis of the contribution that each particular material and manufacturing operation made and the assignment of incremental costs of materials and manufacturing operations for a given part to the function which it provided. A portion of the cost of galvanizing a part might be assigned to the function "resist corrosion" and part to "improve appearance"; the cost of drilling holes might be assigned to the function "allow attachment". (In the case of a slotted hole, the cost to elongate the hole would be attributed to a function such as "adjust alignment".) The VE study team identifies alternative ways to perform the various functions, usually concentrating on the ones with highest cost.

In the early 1960s, Charles W. Bytheway created the Function Analysis system technique (FAST), a structure within which to diagram logically the relationship of Miles' functions. In the 1970s, Professor Thomas J. Snodgrass of the University of Wisconsin contributed a system to collect user and customer attitudes about a VA/VE study object and to rate their importance, assign the importance ratings to functions, and then compare the importance of a function to the cost of the function to identify a value mismatch — a function for which the cost, whether high or low, was mismatched with the customer's perception of its importance. The team concentrates on improving the mismatched functions.

Recently, I led a VE study for a new county jail in California. A significant amount of floor area was consumed by a property storage room. Construction costs allocated to the functional performance requirements of "receive inmates", "classify inmates", and "release inmates" totaled over $3 million, 10.5% of estimated construction costs. Armed with this information, the jail commander decided to eliminate inmate property storage in this new detention-only facility, as the primary function of the jail was to manage sentenced inmates. Release and new inmate intake will be done at the existing jail and inmate property will be stored there.

First costs savings were $500,000. Life-cycle cost savings from eliminating the property storage center staff position and avoiding property storage equipment maintenance accounted for another $1.3 million. Function Analysis had provided the rationale for a decision which up until the time of the study was only a distant thought.

A thorough Function Analysis reviews closely the relationship of three things: function, function cost, and function importance or worth to the customer. The result is targets of opportunity for overall project value improvement, not just cost reduction. This core VE technique can be used at any project stage, from site selection through construction, to help the project team identify and clarify essential project performance requirements. Studies very early in project conception may not require a rigorous assignment of project costs to functions. The process of defining functions and building a function diagram builds understanding of the project.

During a 1-day study, we helped the CEO and top executives of a health-care plan management company use Function Analysis to review, evaluate, and decide among three existing buildings being considered as new locations for the company. While building the function diagram, a breakthrough occurred when the team agreed that their single reason for moving was not to improve operations, but rather to increase sales by providing the firm with a better image to prospective clients who visit their offices.

After reviewing the three potential locations in this new light, the team concluded that none was appropriate to their main requirement for moving. They sought other buildings that could present an image to prospective clients that better reflected their professionalism and capabilities. The design architect was part of the study. The completed function diagram clarified and documented for the architect the client's project performance criteria for the design of the new offices. Assigning functions to activities (claims processing, reception, customer service, etc.) was the initial step in developing design criteria for the spaces in which these activities would occur.

Function Analysis and function diagramming as stand-alone techniques afford the project manager a powerful consensus-building and analytical tool during project planning. During design, assigning costs to functions puts a price tag on program decisions, allowing the customer to answer the question, "Am I getting the result I want for the money I am spending?"

Adhere to the Value Engineering job plan

"Ready, fire, aim!" is not the most effective sequence if the mission is to hit the target with maximum effect using the minimum amount of ammunition. The VE job plan (see Figure B.1) establishes a sequential order of activities, in measured blocks of time, for the VE study team to follow. Each block contains specific tasks to be completed leading toward a desired result. The time allotted to each job plan step depends upon the scope of the project studied.

Following the VE job plan gives the project manager the control necessary to limit the duration of the study while ensuring successful results. The job plan is a flexible tool whose order must be preserved, but its duration can be extended or compressed to fit the project scope. The activities within each step also vary depending upon when in project development the study is carried out. I usually recommend a 3-day team study at the pre-design phase. A longer study, up to 6 or 7 days of team meetings, with individual team-member assignments to

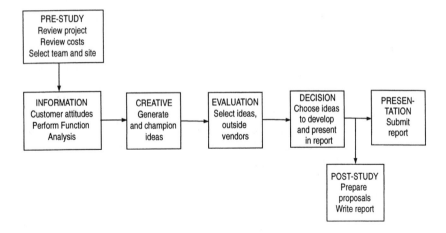

Figure B.1   Value Engineering job plan.

be completed between some of the meetings, is effective during concept design or later.

It is normal to think that the VE team spends most of its time in the creative phase, generating alternative ideas. In reality, the creative phase is the shortest phase of the VE job plan. The most time-consuming and most important VE study activity is a thorough Function Analysis performed in the information phase. I compare the time the VE team spends performing Function Analysis with the time my son spent in high school doing his homework. He learned there was a direct correlation between the time spent on his homework and increasing his grade point.

The information phase is the VE team's homework, and Function Analysis is the primary lesson. In the 2 to 3 days the team may take to work through it, team members become primed to generate collectively and individually by function hundreds of ideas in a short period of time. As a facilitator, my major problem can be halting the creative phase. For example, in a training workshop last year, the lowest number of ideas generated by a team was 198, while the highest was nearly 300 — in only 4 hours! This profusion of ideas provides the seeds from which project value-enhancing changes are grown through thoughtful evaluation, careful development, and owner buy-in.

Conclusion

Modern Value Engineering employs all of the characteristics of proactive, customer-driven processes increasingly being used in project design. A multi-disciplined team of stakeholders working together to understand and accomplish project objectives is the foundation of partnering. Thorough Function Analysis identifies customer quality objectives and the cost to achieve them. It helps the customer to articulate required project performance and acceptable cost to achieve it and then concentrates the entire project team's energy on making it happen. The VE job plan is a framework within which to conduct meetings for the project team to focus on identifying project mission, objectives, and performance and then to work together to achieve them. Modern Value Engineering is a powerful tool to help the project manager work with the client and project team to achieve a quality project at affordable costs.

Appendix C

ISO 9000*

ISO 9000 and ANSI/ASQ Q90 series quality standards guide quality on a worldwide basis. Knowledge of these standards and their certification procedures is essential to contractors and design professionals in order to remain competitive in a global market-driven economy. Customers and suppliers have grown to expect quality efforts to comply with these standards. Failure to do so may eventually cause a company to lose its competitive edge.

ISO 9000 and ANSI/ASQ Q90 standards are made up of five categories. A company seeks certification under the one that is appropriate to its position. The impact of ISO 9000 and ANSI/ASQ Q90 standards is well understood by companies who are competing in the product markets of the world. The significance of ISO 9000 and ANSI/ASQ Q90 standards on the economics and competitiveness of service industries such as construction is less well understood.

Introduction

The International Organization for Standardization (ISO) is located in Geneva, Switzerland. ISO develops and promotes common standards on a worldwide basis. ISO 9000 is the generic name for the series of quality standards adopted in 1987 by the European Economic Community, now the European Economic Union. ISO 9000 as a quality standard is a key element of the international standards for quality management and quality assurance. A company seeks certification under one of the five categories, depending on their individual circumstances.

The series is not a set of product standards, nor is it specific to any one industry. Quality assurance and quality management standards

---

* Written by Jeffery Lew, who is on the faculty of Purdue University Department of Building Construction in West Lafayette, IN (telephone 765-494-2459).

refer to quality system elements that are to be implemented, not the means for implementing them. Quality assurance and quality management system standards are complementary to the standards of a product which affect the functionality of the product or service. Quality assurance and quality management standards refer to quality system elements that are to be implemented, not the means for implementing them. The series is generic and, when used with the appropriate industry-specific quality assurance guidelines, builds a strong foundation for a quality system.

The ISO has over 90 member countries, with the American National Standards Institute (ANSI) representing the U.S., whose participation in ISO is carried out by ANSI-sponsored Technical Advisory Groups. The ISO 9000 series of quality standards was adopted, word for word, by ANSI as the ANSI/ASQ Q90 series.

The five ISO 9000 & ANSI/ASQ Q90 categories are briefly described below:

1. The ISO 9000 and ANSI/ASQ Q90 category is the first in the series of five. It defines the five key quality terms in the standard and is advisory in nature in that it provides guidelines and is the road map for selecting and using the other standards in the series.

2. The ISO 9001 and ANSI/ASQ Q91 category consists of 20 sections applying to firms involved in both the design and the manufacture or production of products or services. It specifies a model for use when a contract between two parties requires the demonstration of a supplier's capability to design, produce, install, and service a product. This category carries the most stringent requirements.

3. The ISO 9002 and ANSI/ASQ Q92 category consists of 18 sections which apply to firms involved only in manufacturing and is more oriented to quality assurance for production and installation than is ISO 9001. Requirements for this category are not as stringent as for 9001.

4. The ISO 9003 and ANSI/ASQ Q93 category covers companies which are involved in distribution of manufactured products or services only. It serves as a model for quality assurance in final inspection and testing. Requirements for 9003 certification are the least stringent of the series.

5. The ISO 9004 and ANSI/ASQ Q94 category is again advisory in nature and offers quality management and quality system guidelines on the elements of quality management systems and quality systems to determine the extent to which each is applicable.

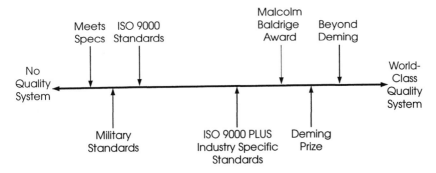

Figure C.1   ISO 9000 certification compared to other standards. (From the Construction Industry Institute [CII] Forum. For more information regarding the CII, call 512-471-4319.)

These series of standards are not rigid and inflexible. The ISO 9000 series is flexible and is periodically reviewed to take into account methodologies that prevent defects, such as statistical process control for projects and the employment of self-managing work teams.

The ISO 9000 and ANSI/ASQ Q90 standards need to be understood as the minimum required standard for an acceptable quality system. This must not be taken as the final mark of excellence. Many sets of quality standards and awards are currently being used. Figure C.1 indicates a relative ranking for the more common quality system benchmarks. The ISO 9000 and ANSI/ASQ Q90 standards should be used as a minimum standard upon which to build a quality system on a strong foundation. The system can then be further developed to achieve a desired level of excellence.

Background and development of ISO 9000

ISO 9000 quality system standards describe a comprehensive but basic set of quality assurance standards that have many elements common with U.S. Department of Defense (DOD) quality specifications used for the military. These DOD standards were adopted by the North Atlantic Treaty Organization (NATO), then broadened and strengthened by the British. Finally, the evolutionary process of these quality standards was completed with their adoption by ISO in 1987 to facilitate free trade within the European Community. In 1993, DOD replaced its MIL specifications with ISO 9000.

The American National Standards Institute (ANSI) with the American Society for Quality (ASQ), rather than independently revising and extending their existing quality systems standards, elected to join with other nations in adopting standards fully consistent with ISO 9000. Standards adopted as ANSI/ASQ Q90 through Q94 were written to be equivalent technically to the ISO 9000 through 9004 series. The benefit of this procedure was to incorporate customary American language usage and spelling into the standard. Thus, for all practical purposes, the ISO 9000 series and the ANSI/ASQ Q90 series have the same meaning.

Description of ISO 9000

Certification

Certification of a company under one of three categories of 9001, 9002, or 9003 is granted by a certifying organization described in the next paragraph. Certification to ISO 9000 confirms that a company's practices and procedures are consistent with ISO 9000 standards. In other words, a firm is certified that it actually does what it says it does. Certification is required for each facility (location or office) within a particular company. The certifying procedure is carried out in three steps:

1. The certifying organization compares the applicant's quality manual with the 9000 standard and verifies that procedures and systems are consistent with the manual.
2. The auditors of the certifying organization conduct an initial audit of the applicant's facility to verify that practices and systems comply with the procedures in the manual. Suppliers, vendors, and customers are contacted to confirm consistency with the manual.
3. The certifying organization issues certification after successful completion of the above steps. Certification desired under ISO 9000 and ANSI/ASQ Q90 can be obtained by asking the certifying organization to include both on the certification paperwork.

The typical steps required to achieve certification of a quality system are outlined in Figure C.2.

Certifying organizations

Nearly 40 organizations (many European) are presently authorized by agencies such as the National Accreditation Council for Certification Bodies of the United Kingdom to offer ISO 9000 certification. In addition, ANSI/ASQ and other organizations are establishing processes for direct domestic certifying organizations and accreditation of registrars.

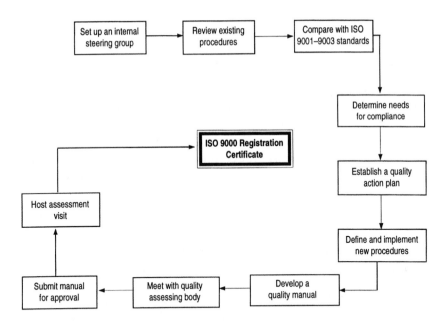

Figure C.2   Steps for ISO certification.

A partial list of Quality System Registrars is provided at the end of this appendix. Registration information may be obtained from Perry Johnson, Inc., or any firm listed at the end of this appendix. A complete list of registrars may be obtained from CEEM Information Services (see the end of this appendix).

Certification requirements

The principal, unifying requirement for certification is documentation. The critical documentation item is the firm's quality manual. The manual must be developed in a prescribed format with the appropriate language and include documentation systems. In summary, a firm must be able to document, plan, implement, and evaluate an ISO 9000 quality system.

The manual serves as a tool to incorporate the ISO 9000 series of quality standards into an organization. By covering the various sections of the ISO 9000 standard in a clear and concise fashion, documentation of how a firm meets the standard is provided.

It should be stated that it is not difficult to obtain and retain ISO 9000 certification. ISO 9000 is considered less stringent than many quality requirements presently in effect. ISO 9000 is considerably less demanding than the requirements of the Malcolm Baldrige National

Quality Award. However, it is possible and likely that certification requirements will become increasingly more difficult with time.

Management should note that all employees of a facility are involved in the ISO 9000 implementation process. This is true for any quality process. The certifying auditors can talk randomly with personnel throughout an applicant's organization, including outside suppliers and customers. Therefore, all employees must understand the requirements of the ISO process in order to obtain successful certification. While training is not required for ISO certification, it is certainly needed in a practical sense and should be up and running before final certification.

Quality systems elements

One should understand that ISO 9000 is intended to pertain to all businesses, including the service (or construction) industry, as well as manufacturing. The system is, of course, generic and as such is philosophical, but not necessarily specific, in nature. The quality system elements required under ISO 9000 and ANSI/ASQ Q90 standards are listed below:

> Contract review
> Control of non-conforming product*
> Corrective action
> Design control and servicing**
> Document control*
> Handling, storage, packaging, and delivery*
> Inspection and testing*
> Inspection and test status*
> Inspection, measuring, and test equipment*
> Internal quality audits
> Management responsibility
> Process control
> Product identification and traceability*
> Purchaser supplied product
> Purchasing
> Quality records*
> Quality system*
> Statistical techniques
> Training

---

* ISO 9001 only.
** ISO 9003 includes only these elements.

Time and cost

It is difficult to assess the time that it takes to obtain certification. This is due to the fact that the time required depends on the applicant's quality system status at the time the registration process begins. Again, the key point here is documentation. If a firm's documentation coincides with the model that auditors expect to see, the certification process can take as little as 6 months. Firms starting from scratch without any formal quality systems can take 18 months to 3 years to obtain certification. The cost of certification can be $5000 to $10,000, along with 1000 to 1500 hours for audit preparation. The actual audit takes 2 to 4 days to conduct.

Application to design and construction

Design and construction firms and contractors are concerned with the institutionalization of quality systems. The procedure for applying for certification is given in the following paragraphs as a typical example of how to go about obtaining certification.

1. A firm must first understand what the ISO 9000 and ANSI/ASQ Q90 standards are, along with the definition of the key elements of Total Quality Management (TQM).
2. An applicant must understand the ISO 9000 and ANSI/ASQ Q90 series standards, as described previously in the introduction to this appendix.
3. The applicant determines which of the ISO 9000 and ANSI/ASQ Q90 standards apply best to their firm. Most design and construction firms and contractors should seek certification under ISO 9001. A case can be made for 9003, but these firms would have to demonstrate that their final inspection would catch all errors and that proper corrective measures could be provided.
4. The registration and certification of a firm's facilities is normally made one location at a time, e.g., one certification for each design office or contractor's office. Registration can also be made for a single system at a particular location. The quality system must provide an end product (or service) to an external customer, such as a set of plans for a waste-water treatment plant. In summary, registration and certification of an entire firm requires that each office or each construction division go through the application process and be audited.
5. The preparation of the manual described earlier must contain a description of everything that a company does and should provide

an exact duplication of what is being done. There are two types of manuals required for the audit.

a. The policy manual, or who does what, includes policy statements, describes responsibilities, and denotes the authority of personnel. Also included is the written commitment of company management to include their involvement and leadership.

b. The systems manual, or how things are done, includes detailed descriptions of all systems in a company. This means comprehensive and detailed flow charts of the work-flow process with responsibilities shown for work and quality. However, note that only issues relating to assuring quality should be addressed in this manual. Companies that perform engineering and engineering design need to include these functions in quality documentation systems.

Conclusion

Although ISO 9000 and ANSI/ASQ Q90 series certifications are not required domestically in design and construction, they are a necessity for overseas operations. Overseas companies are becoming involved in certification, and domestic companies also need to become involved.

One area in which ISO 9000 and ANSI/ASQ Q90 may come into play in the near future is international work in Canada and Mexico, as the North American Free Trade Act (NAFTA) is implemented. A number of U.S. contractors are looking for work in Mexico and Canada, where ISO 9000 and Q90 will help them manage and work with local suppliers and subcontractors. Design and construction firms need to understand the certification requirements, as they are used by many owners, suppliers, and vendors (in other words, the customers of the construction industry.)

Quality System Registrars (partial list)

ABS Quality Evaluation, Inc.
Robert C. Sutton, President
263 North Belt East
Houston, TX 77060
Telephone (713) 873-9400
FAX (713) 874-9564

AT&T's Quality Registrar
John Malinauskas
650 Liberty Avenue
Union, NJ 07083
Telephone (908) 851-3058
Toll-free (800) 521-3399
FAX (908) 851-3360

Bureau Veritas Quality (International NA), Inc.
Greg Swan
509 North Main Street
Jamestown, NY 14701
Telephone (716) 484-9002
FAX (716) 484-9003

Perry Johnson, Inc.
Barbara Schreier
3000 Town Center, Suite 2960
Southfield, MI 48075
Telephone (313) 356-4410
FAX (313) 356-4230

Quality Systems Registrars, Inc.
Scott Kleckner
1555 Naperville/Wheaton Road, Suite 206
Naperville, IL 60563
Telephone (630)-778-0120
FAX (630) 778-0122

Steel-Related Industries Quality Systems Registrar
Peter B. Lake
200 Corporate Drive, Suite 450
Wexford, PA 15090
Telephone (412) 935-2844
FAX (412) 935-6825

Quality System Registrar Directory
CEEM Information Services
10521 Braddock Road
Fairfax, VA 22032
Telephone (800) 745-5565
FAX (703) 250-5313

Appendix D

Selected list of computer software for design firm project management, financial accounting, and scheduling

Project management and financial accounting

Axium Systems, Inc.
419 Canyon Avenue, #300
Fort Collins, CO 80521
(970) 224-9644

BST Consultants, Inc.
5925 Benjamin Ctr., #110
Tampa, FL 33634
(813) 886-3300

Harper & Shuman, Inc.
68 Moulton Street
Cambridge, MA 02138
(617) 492-4410

Semaphore, Inc.
3 East 28th Street, 11th Floor
New York, NY 10016
(212) 545-7300

Wind-2, Inc.
1901 Sharp Point Drive, #A
Fort Collins, CO 80525
(303) 482-7145

Scheduling

Allegro Systems
5690 DTC Blvd., #260
Englewood, CO 80111
(303) 850-9854

Microsoft Project
(available from most computer software stores)

Primavera Systems, Inc.
Two Bala Plaza
Bala Cynwyd, PA 19004
(610) 660-5843

Index

Index

A

account managers   13
activity, definition of   93
American Consulting Engineers
    Council (ACEC)   43, 149
American Institute of Architects
    (AIA), survey by   89
American National Standards
    Institute (ANSI)   204, 205
American Society for Personnel
    Administration (ASPA)   48
American Society for Quality
    (ASQ)   205
American Society for Training and
    Development (ASTD)   48
American Society of Civil Engineers
    (ASCE)   66
American Society of Heating,
    Refrigerating, and Air-
    Conditioning Engineers
    (ASHRAE)   66
ANSI/ASQ Q90   203–211. *See also*
    ISO 9000 certification
    categories of   204
Association for Project Managers
    (APM), survey by   4, 17, 27,
    28, 36
Association of Engineering Firms
    Practicing in the Geosciences
    (ASFE)   149

B

bar charts   92. *See also* Gantt charts
billing checklist   126
budgeting of projects   75–89
budgets   24, 26, 35, 75–89
    development of   79
    sample form   78

C

CADD   22, 40, 59, 73, 116, 122, 148,
    157, 162–163, 164–165
    evaluating need for   158–159
    strategies for use   159–163
    using effectively   159
    willingness to use   159
career tracks   22, 47
cartooning   169
certificate of insurance   73
change orders   73, 127, 128–129
chargeable rates   58, 110. *See also*
    reimbursable expenses
check-sets   137
checklists   136
clients
    communication with   61, 142
    expectations of   62–64
        clear point of contact   62–63
        communication   142
        cost control   63